Bezabih Yimer

Alternative energy sources to combat climate change

Biogas production using cost effective material

Anchor Academic
Publishing

Yimer, Bezabih: Alternative energy sources to combat climate change. Biogas production using cost effective material. Hamburg, Anchor Academic Publishing 2013

Buch-ISBN: 978-3-95489-127-6
PDF-eBook-ISBN: 978-3-95489-627-1
Druck/Herstellung: Anchor Academic Publishing, Hamburg, 2013

Bibliografische Information der Deutschen Nationalbibliothek:
Die Deutsche Nationalbibliothek verzeichnet diese Publikation in der Deutschen Nationalbibliografie; detaillierte bibliografische Daten sind im Internet über http://dnb.d-nb.de abrufbar.

All rights reserved. This publication may not be reproduced, stored in a retrieval system or transmitted, in any form or by any means, electronic, mechanical, photocopying, recording or otherwise, without the prior permission of the publishers.

Das Werk einschließlich aller seiner Teile ist urheberrechtlich geschützt. Jede Verwertung außerhalb der Grenzen des Urheberrechtsgesetzes ist ohne Zustimmung des Verlages unzulässig und strafbar. Dies gilt insbesondere für Vervielfältigungen, Übersetzungen, Mikroverfilmungen und die Einspeicherung und Bearbeitung in elektronischen Systemen.

Die Wiedergabe von Gebrauchsnamen, Handelsnamen, Warenbezeichnungen usw. in diesem Werk berechtigt auch ohne besondere Kennzeichnung nicht zu der Annahme, dass solche Namen im Sinne der Warenzeichen- und Markenschutz-Gesetzgebung als frei zu betrachten wären und daher von jedermann benutzt werden dürften.

Die Informationen in diesem Werk wurden mit Sorgfalt erarbeitet. Dennoch können Fehler nicht vollständig ausgeschlossen werden und der Diplomica Verlag, die Autoren oder Übersetzer übernehmen keine juristische Verantwortung oder irgendeine Haftung für evtl. verbliebene fehlerhafte Angaben und deren Folgen.

Alle Rechte vorbehalten

© Anchor Academic Publishing, Imprint der Diplomica Verlag GmbH
Hermannstal 119k, 22119 Hamburg
http://www.diplomica-verlag.de, Hamburg 2013
Printed in Germany

Acknowledgements

I would like to express my sincere and deepest gratitude to my brother Mr. Seid Yimer (M.Sc in Chemical Engineering, Wollo University, Ethiopia) for his intellectual advice and encouragement for the completion of the book.

My special gratitude is due to my beloved wife Mrs.Serkalem Moges and entire families for their unlimited encouragement and technical supports. Above all, my special thanks direct to the Almighty God, in offering me all the patience and endurance during the preparation of this book.

Table of contents

ABSTRACT 1

1 Introduction 3
   1.1 Background 3
   1.2 Problem Statement 4
   1.3 Purpose of the Study 4
   1.4 Hypothesis 5
   1.5 Objectives of the Study 5
      1.5.1 General objective 5
      1.5.2 Specific objectives 5

2 Literature Review 6
   2.1 Fuel Consumption in Ethiopia 6
   2.2 Biomass and Biogas Energy Technologies in Ethiopia 7
   2.3 Theory of Biogas Technology 8
   2.4 Benefits of Low- Cost Plastic Biodigester Technology 8
      2.4.1 Environmental Benefits of Biogas Technology 10
      2.4.2 Social Benefits of Biogas Technology 10
      2.4.3 Economic Advantages of Plastic Biogas Technology 10
      2.4.4 Beneficiaries of Biogas Development 11
   2.5 Input Materials for Bio- Gas Production 11
   2.6 Biogas Production Processes 12
   2.7 Theory of Biogas Burner 13
   2.8 The Slurry after Digestion 13
   2.9 Measurement of Biogas Production 14
   2.10 Designing of Digesters 14
   2.11 Working of Fixed-Dome Biogas Plant 14
   2.12 Selection and Layout of Pipeline and Biogas Accessories 15
   2.13 Transfer of the Plastic Film Biodigester Technology 15
   2.14 Promotion of Fixed and Floating Dome Biogas Plant 16
   2.15 Economic Evaluations of Biogas Plants 16
   2.16 LDPE Geomembrane Plastic 16
   2.17 Theory of Environmental Impact Assessment (EIA) 18

3 Materials and Methods 19

3.1. Description of the Study Area 19

3.1.1 Location 19

3.1.2 Socio-economic activity 19

3.1.3 Climate 20

3.1.4 Land Use 20

3.1.5 Livestock Population 21

3.2 Experimental Design and Layout 21

3.3 Geomembrane Plastic Construction methodology 23

3.4 Data Collection Procedures 24

3.4.1. Input to the Digesters 24

3.4.2 Measurement of Gas Production 25

3.4.3 Temperature of the Air and Slurry 25

3.4.4 Total-Solids (DM) Content 25

3.4.5 The Organic Dry Matter (ODM) 25

3.4.6 pH of the Fresh Cow Dung and Digested Slurry 26

3.4.7 Quality of Output Slurry 26

3.4.8 The Efficiency of Biodigesters 26

3.4.9 Social Aspect of Biomass and Biogas Technologies 27

3.4.10 The Economic Viability of a Plastic and Fixed Dome Biogas Plant 27

3.4.11 The Environmental Impact of the Plastic Biogas Plant 27

3.5 Statistical Analysis 27

4 Result and Discussion 29

4.1 Operation of Plastic Biodigester 29

4.2 Biogas production 29

4.3 Temperature of the Air and Slurry 31

4.4 Characteristics of Bio-digested Slurry (Effluent) and the Influent 32

4.5 Characteristics of Total-N in the Slurry and Influent 33

4.6 Characteristics of Organic Matter in the Slurry and Substrate 36

4.7 Characteristics of pH of Fermented Slurry 37

4.8 Efficiency of the Biodigesters 38

4.9 Economic Evaluations 38

4.9.1 Market Price of Inputs 39

4.9.2 Market price of outputs 39

 4.9.3 Cost-Benefit Analysis of Biogas Plants 41
4.10 Social aspect of biogas technology 46
 4.10.1 Income generation through increased crop production 46
 4.10.2 Income Generation through Cost Saving 46
 4.10.3 Perceptions of Habru Woreda People regarding the use of Biomass & Biogas Technology 47
4.11 Technological aspect of geomembrane plastic biodigester 47
 4.11.1 Sustainability 47
 4.11.2 Simple technology 48
 4.11.3 Demand driven 48
 4.11.4 Replicability 48
4.12 Technical problems with the geomembrane plastic digester 49
4.13 Environmental Impact Assessment of the Plastic Biodigester 50
 4.13.1 Reduction of green house gas emissions 50
 4.13.2 Reduction of rate of deforestation 52

5 Conclusions and Recommendation 53
 5.1 Conclusions 53
 5.2 Recommendations 55

References 56
Appendix 62
 Appendix 1 62
 Appendix 2 62
  Limitations 62
 Appendix 3 66
 Appendix 4 68
 Appendix 5 68
 Appendix 6 69
 Appendix 7 71
 Appendix 8 72
 Appendix 9 72
 Appendix 10 73
 Appendix 11 74

List of Tables 75
List of figures 75
Acronyms 77

ABSTRACT

The study was conducted in North Wollo, Mersa-Chekorsa village, Ethiopia in 2006/2007, where animal dung for biogas production is available. The overall objective of the study was to introduce economically feasible, technically acceptable and environmentally friendly biogas plant to the farming community and other potential users in Ethiopia. The research was carried on two types of biogas plants of $3m^3$ capacity (1) geo-membrane plastic (two single and two double layered) biogas plants constructed below and above the ground surface and (2) fixed-dome biogas plant. Each bio-digesters was fed with a mixture of 75Kg of cow-dung and 75Kg pure water at equal volume and proportion. Amount of gas and slurry were measured using calibrated biogas burner and weight balance respectively. The quality of the slurry (i.e. total-N and organic matter content) were analyzed in the laboratory using Kjeldahl and ash method respectively. The bio-digesters were compared after gas has completely produced at the end of 40 days of fermentation with respect to amount of gas and slurry produced, quality of slurry in terms of total-N and organic matter content. Economic analysis of the bio-digesters was carried out using cost-benefit analysis. The social aspect of using biomass and biogas technologies and environmental impact assessment of the new geo-membrane plastic biogas technology was also assessed. The emissions of CO_2 and CH_4 were computed by measuring the production of biogas in the two models of bio-digester. Fermented slurry contained larger nitrogen content than fresh cow dung in both models of bio-digester. The geo-membrane plastic biogas plant gave higher net benefit than fixed-dome biogas plant. So, from this, investment on geo-membrane plastic bio-digester is economically feasible. Environmental impact assessment of the technology was studied and found that 360.04 m^3 of CO_2 and 600.06 m^3 CH_4 was prevented from emitting in to the atmosphere and save 0.562 hectare of forest per year. Generally, it was found that, the geo-membrane cylindrical film bio-digester technology was found cheap and simple way to produce gas in the study area and it was recommended to introduce the technology into the rural areas having even and high temperature which is similar to the study area more preferably to an area having mean daily temperature greater than $20\,^{O}C$.

Key words: - Geo-membrane Plastic bio-digester, fixed-dome bio-digester, biogas production, quality of fermented slurry, environmental impact assessment and economical feasibility

1 Introduction

1.1 Background

Farming is the major rural activity in Ethiopia, i.e. agriculture supplies 51.8% of the gross domestic product and 90% of the export earnings of Ethiopia and 86% of the population is engaged in agriculture (CSA, 1999 cited in Paulos, 2004). Dependency from biomass such as fuel wood, charcoal, dried cow dung cake and crop residue in Ethiopia amounts to 95% and half of the biomass is used for baking injera (Benjamin, 2004). When all forest uses are included, the deforestation rate in Ethiopia is around 1.1% per year (Wikipedia, 2006). According to North Wollo Agricultural and Rural development office (2007), the forest cover of North Wollo and Habru district is 37,183.58 hectare and 1614 hectare respectively.

Fuel is in very short supply in Ethiopia and throughout most of Africa. Where conditions still permit, wood is commonly used as fuel, but in many rural and urban areas, dried cow dung is a major source of fuel for cooking (Benjamin, 2004). Wood burns at 5-8 % efficiency and cow dung at 60% of that of wood (UNESCO, 1982).

According to FAO (2000), the combustion of fossil fuels has caused serious air pollution problems, likewise the excessive consumption of fire wood results in deforestation on a large scale. The deteriorating forest cover in Ethiopia due to deforestation caused the recurrent drought and famine (FAO, 2000). IUCN (1990), estimated that high forests covered 16% of the land area of Ethiopia in the early 1950s, 3.6% in the early 1980s and 2.7% in 1989.It is estimated that these resources are vanishing at an alarming rate, estimated at 150,000 to 200,000 hectares per year (EFAP, 1994).Therefore, deforestation had caused and continuous to cause environmental degradation in the form of land degradation, water resource deterioration and lose of bio-diversity.

Biogas digestion was introduced into developing countries as a low - cost alternative source of energy to partially alleviate the problem of acute energy shortage for households, reduces deforestation and soil erosion, avoids scarcity of firewood, benefits environment globally and provides excellent fertilizer, there by increases crop production (Vandana,2004).

However, few farmers used biogas in practice in Ethiopia (Yacob, 2000). So, to solve the problem of biogas technology dissemination, it is very important to study on alternative biogas plants constructed using a different material and design.

Thus, the economic assessments of the new and previous model digesters consider adequately the costs of traditional alternative sources of fuel, and the benefits of using biogas and output slurry for cooking and as fertilizer. It is also essential to assess the environmental impact of both models of biogas plants by computing the replacement of conventional fuels such as reduction of fuel wood, cow dung and chemical fertilizer with the use of the biodigesters.

1.2 Problem Statement

Fuel wood and charcoal are the primary sources of energy for Ethiopia's rural population. According to MOA (2000), on average each rural household spends ten hours per week searching for fuel wood. Females & children are engaged to search fire wood for about 5-6 hours journey (MOA, 2000). Consumption of charcoal and fuel wood is a serious factor in deforestation, environmental degradation, air pollution and carbon dioxide emissions (Paulos, 2004). In Ethiopia with the current economic status almost the majority of family spent 20-30% of their monthly income for purchasing fire wood (MOA, 2000).

Given the scarcity of firewood and the prevalent use of dung for cooking, biogas plants would appear to be an appropriate means of reducing the current usage of these renewable energy resources. The biogas technology used almost exclusively today in Ethiopia, i.e. fixed-dome and floating –drum biogas plants became an obstacle to the rapid diffusion of biogas technology, because it takes a relatively long time (3-5 week) to build a plant and high initial cost of investment, shortage of adequate skilled person who can undertake construction & installation of the plant and transportation problem of the prefabricated steel drum from the urban areas to the interior rural regions of Ethiopia (Yacob,2000).

Considering the problem of biogas technology dissemination with the existing biogas plants, the study was conducted on alternative biogas plants constructed using geomembrane plastic in cylindrical shape.

Therefore, in order to introduce geomembrane plastic biogas plant, comparisons of gas and slurry yield and economic feasibility analysis with the fixed-dome biodigester should be done and accordingly, the outcome of the study may have some contribution to set remedy to the problem.

1.3 Purpose of the Study

The purpose of the study is to ensure the wide use of biogas plants by the farming community, reducing the workload of women by using low-cost biodigester, to achieve ecological stability, namely burning biogas instead of firewood, and not using cattle dung cake and other agricultural waste directly as fuel, but using it in biogas plants to produce both fuel and fertilizer.

The output of the study provides economically feasible, technically efficient and environmentally sound geomembrane plastic biogas plant.

1.4 Hypothesis

1. Geomembrane plastic biogas plant could provide a cheaper and higher amount of gas and slurry than fixed-dome biogas plant.
2. The negative impact of conventional fuels on the environment can be reduced by replacing their use with biogas plant.

1.5 Objectives of the Study

1.5.1 General objective

The main objective of the present investigation is to introduce economically feasible, technically acceptable and environmentally friendly biogas plant to the farming community and other potential users in Ethiopia.

1.5.2 Specific objectives

1. To measure the quantity of gas, quality and quantity of slurry produced from geomembrane plastic and fixed dome biogas plant.
2. To evaluate the economic feasibility of geomembrane plastic and fixed dome biogas plant.
3. To evaluate the environmental impact of the geomembrane plastic and fixed-dome biogas plants.
4. To document the utilization of biomass and biogas technologies by the surrounding farming community.

2 Literature Review

2.1 Fuel Consumption in Ethiopia

About 95% of the total energy consumption in Ethiopia is provided by wood, dung, charcoal and crop residues. Biomass consumption for fuel in 1980 was about 24 million m^3 of firewood (mainly *Acacia* spp.and *Eucalyptus* spp), 7 million tones of dung, 170,000 tones of charcoal and 6 million tones of crop residues (FaWCDA, 1982). Half of the biomass was used for baking injera and the use of fossil fuels (petroleum products) account only for 5% of primary energy in Ethiopia, but they cost nearly 50% of export earnings (Benjamin, 2004).

At present more than 90% of the domestic supplies of industrial wood and firewood comes from the natural forests which are the main sources of wood products. Fuel wood accounts for the bulk of the wood used, and is the predominantly preferred domestic fuel in both rural and urban areas. The projected demand for fuel wood based on assumed per capita requirement is on the increase and is expected to be over 100 million m^3 by 2020. On the other hand, the projected supply from all sources is expected to be only 9 million m^3 which is far below the demand (FaWCDA, 1982). Thus, more efficient ways of firewood utilization need to be investigated and substantial savings in firewood consumption could be achieved by the use of biogas and different cooking methods.

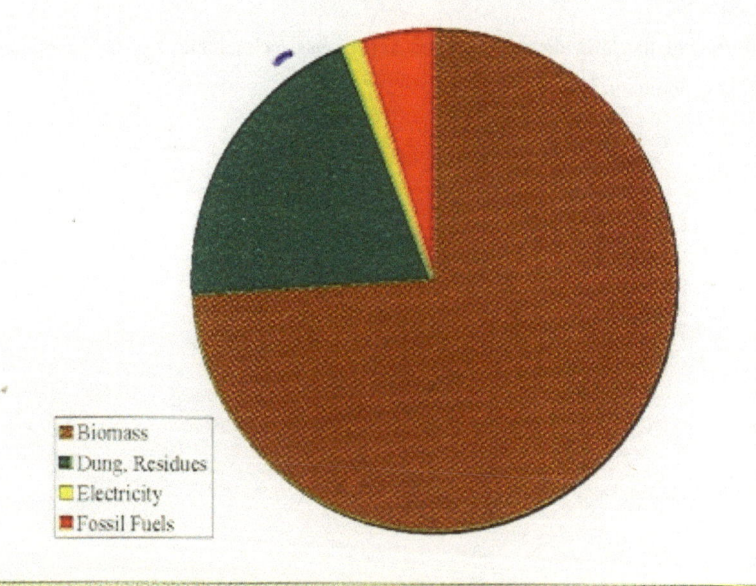

Figure 1: Ethiopia's Primary Energy Shares, Benjamin Jargstorf (2004)

2.2 Biomass and Biogas Energy Technologies in Ethiopia

The energy balance of the country as of 1995/96 reveals that total energy consumption in Ethiopia was estimated at 723 peta joules or about 50 million tones of wood equivalent and is characterized by high dependence on biomass fuels. The contribution of wood fuel alone was about 77 percent of the final consumption and agricultural residue and dung accounted for about 16 percent which means that the share of traditional fuels in the national energy consumption was above 90 percent. Modern energy (petroleum fuel and electricity) accounted only for about 5.5% of total energy consumption, the share of petroleum being about 4.8% and that of electricity being 0.7%. About 90 percent of the overall energy consumption of the country is that of households out of which the share of urban households is only 6 %. Rural households almost entirely rely on the traditional fuels where as the share of modern fuels in urban households' consumption was about 20 percent. So, the extent of dependence on traditional fuels is very high
(Samuel, 2007).

Figure 2: Fuel Wood Carriers for Fuel Consumption from the Forest, Agent of Deforestation, Benjamin Jargstorf (2004).

Figure 3: Dry Cattle Dung Cakes used for Fuel in Ethiopia, Benjamin Jargstorf (2004).

Thus, it makes sense to 'modernize' biomass it self and convert it from its solid form to more convenient forms such as gases, liquids or electricity. The most promising technology that can convert biomass into such clean, efficient gas is biogas plant.

2.3 Theory of Biogas Technology

Biogas technology refers to the production of a combustible gas (called bio- gas) and a value added fertilizer (Called sludge) by the anaerobic fermentation of organic materials under certain controlled conditions of temperature, pH, HRT, C:N ratio etc. A typical bio-gas plant consists of input unit for feeding the fermentable mixture, a digester where anaerobic fermentation takes place, a gas holder for collecting the bio-gas and to cut off air to the gas outlet pipe and out put unit for removal of fermented slurry (Vandana, 2004). The plant operates on the principle that when dung and other organic materials are fermented in the absence of air, combustible methane gas is produced (Vandana, 2004).

According to Grewal et al.(2000),biogas usually contains 50-65% methane (averaging 60%), 30-40% carbon dioxide (averaging 36%),1-5% of hydrogen,1% nitrogen, 0.1% oxygen, ,0.1% hydrogen sulphide and 0.1% water vapours (H_2O).

2.4 Benefits of Low- Cost Plastic Biodigester Technology

Global level: - Using biogas for cooking reduces the need for fuel wood and charcoal. Studies conducted by a Tanzania local energy NGO indicates that every 8 households clear fell one hectare of forestry each year through charcoal consumption alone (SURUDE,2002). When other causes of forestry destruction are added such as fuel wood, agriculture, construction and mining, deforestation rate in Tanzania is estimated at between 300,000 hectares and 400,000 hectares per year. Studies have further shown that each biogas unit is able to reduce scale of deforestation by 37 hectares per year. Since, it also uses cow dung that would otherwise have degraded, further green house gas emissions are avoided. This is realized by adapting to biogas in place of fuel wood and charcoal for cooking & heating (SURUDE, 2002).

National level:- Bio-gas helps to save foreign currency which is spent on kerosene and chemical fertilizers. Researchers have estimated that 5 lakh bio-gas plants will have 750 million liters of kerosene per year and provide 12 million tones of organic manure. Biogas helps in reducing the need for expensive energy distribution in rural areas. Due to inefficient distribution system almost 20 percent of the powers are lost during transmission. Biogas system would help in preventing the denuding of forests in a careless manner by the villagers for fire wood requirements. Today deforestation being a serious threat to environment in large parts of the country, as it is followed by the danger of soil erosion and several other ecological imbalances (Vandana, 2004).

Local level: - Reduced deforestation helps preserve forests and all of the services they provide, such as biodiversity and maintenance of water quality. In addition, the promotion of agro forestry practices in conjunction with livestock helps protect soil fertility, prevent erosion, and reduce the risk of overgrazing problems often associated with cattle (Duong et al., 2002).

Poverty Alleviation:- Biogas production integrated with cattle raising and farming provides a reliable source of cleaner fuel as well as increased in come and employment opportunities. Therefore, increased incorporation of cattle in to farming methods increases employment opportunities there by stimulating rural economy. The production of biogas also produces slurry that is very effective as a fertilizer. Farmers have effectively used it in agro forestry farming. Studies by Sokoine Agricultural University in Tanzania have shown that the use of this fertilizer helps maintain soil quality over time, there improving crop yields (SURUDE, 2002).

Poverty reduction through improved health:-Respiratory diseases and sometimes deaths caused by indoor pollution as a result of prolonged exposure to smoke from fuel wood and charcoal is avoided when biogas is used for cooking (Vandana, 2004). The utilization of biogas freed the house wives from eye- sore, eye and lung diseases. The use of bio- gas as a domestic fuel can be a thrilling experience for a house wife (SURUDE, 2002).

Reduced drudgery: - Women and children do not have to spend as much time looking for firewood. Cooking with biogas is also faster than with firewood. As a result, the drudgery and workload of women is lessened. Cooking by using a biogas cooker is easy and fast, this has two implications. On one hand it has reduced fuel wood collection and pollution laden cooking tasks on the part of women. On the other hand it has increased gender equity by involving men in domestic chores. Projects that provide direct benefits to woman are usually sustainable (SURUDE, 2002).

Table 1: Work Load before and after Biogas Production

Activity	Hrs./day required to complete the activity before biogas introduction.	Hrs. required to complete activity after biogas introduction.	Hrs. of the household labour per day saved because of the introduction of biogas.
Collection of dung	0.4	0.5	-0.1
Feeding biodigester	0.0	0.5	-0.5
Collecting firewood	4.0	0.5	3.5
Cooking and cleaning	4.2	2.1	2.1
Slurry transport	0.0	0.0	0.0
Total	8.6	3.6	5.0

Source: African Development Foundation Assessment Report in Tanzania (2004).

2.4.1 Environmental Benefits of Biogas Technology

Biogas does not contain toxic carbon monoxide so no danger to health and no offensive odour, reduction in pollution as BOD and COD and facial pathogens are considerably reduced and environment improvement in rural area reduces illness and build up people's health. Besides, in regions where biogas is used to generate electricity, cultural, recreation and spare time study conditions can also be improved. (Duong et al., 2002).

2.4.2 Social Benefits of Biogas Technology

Biogas development brings about social benefits. As the problem of fuel for the farmer's daily use is solved, trees are protected and forests are developed. The protection of trees and increase in vegetation areas can reduce soil erosion and improve ecologic balance. The increase in organic manure can result in using less chemical fertilizer, improving soil and increasing production (UNV, 1983). Therefore, the use of plastic biogas plant saves the time that can be used for wage work, consumption of conventional energy sources for cooking, lighting or cooling and substitution of digested slurry in place of chemical fertilizers and / or financially noticeable increased in crop yields.

2.4.3 Economic Advantages of Plastic Biogas Technology

Production and utilization of biogas are beneficial in many ways. They have both direct and indirect economic benefits. The direct economic benefit of biogas as a fuel, in place of firewood and coal, is a reduction in fuel expenses. Compared with direct burning of stalks; biogas produced from biomass fermentation increases the quality of organic manure which can be sold to production teams, increasing the direct benefit to farmers. Biogas production also has many indirect benefits, which sometimes play a very important role in biogas development. For instance, crop stalks, when no longer burned, may be used as animal fodder, increasing the income from animal husbandry, while still providing raw material for biogas production. Farmers can use the time saved from fire wood collection for additional production, and there by increases their income; fermentation effluent can be used as fodder to raise fish, mushrooms and earth worms, and a proteins fodder for poultry (GTZ, 1989). Therefore, from the use of plastic biogas technology, national energy savings, primarily in the form of wood and charcoal, with the later being valued at market prices or at the cost of reforestation, reduced use of chemical fertilizers, so organic crop yield could be obtained and additionally, foreign currency may be saved due to reduced import of energy and chemical fertilizers could be acquired as an advantage.

2.4.4 Beneficiaries of Biogas Development
Rural farming families benefit most from biogas. Women benefit especially, since biogas reduces their workload, improves their access to income through the sale of milk, and reduce health problems (Lekulel et al., 2002).

2.5 Input Materials for Bio- Gas Production
The following organic matter rich feed stocks are found feasible for their use as input materials for bio-gas production (Vandana, 2004).

Animal Wastes
Cattle dung, urine, goat and poultry droppings, litter, house wastes, fish wastes, leather wastes, sericulture wastes, elephant dung, piggery wastes, etc.
Human wastes
Faces, urine and other wastes emanating from human occupations.
Agricultural wastes
Aquatic and terrestrial weeds, crop residues, spoiled fodder, tobacco wastes, oil cakes, fruit and vegetable processing wastes, cotton and textile wastes, spent coffee and tea wastes.
Waste of aquatic origin
Marine plants, twigs, water hyacinth and water weeds.
Industrial wastes Sugar factory, tannery and paper wastes are industrial wastes that can be used as input material for biogas production. When the cattle dung is used as a feed stock, the

biogas plant is to be filled with homogeneous slurry made from a fresh dung and water in a ratio of 1:1 according to the quantity of input design calculation (Table 2).

Table 2: Quantity of cattle dung Required for Feeding of different Sizes of Bio-gas units

Size of plant (m^3)	Amount of dung required daily (kg)	Appropriate Number of adult cattle heads	
		Local	Cross-breed
1	25	2	1
2	50	4	2
3	75	6	3
4	100	8	4
6	150	12	6
8	200	16	8
10	250	20	10
15	375	30	15
20	500	40	30

Source: Training course of IREP, organized by planning commission and Gandigram Rural Institute (2004).

Table 3: Potential gas Production from Different Feedstock

Type of feedstock (Dung)	Gas yield per kg (C.U.M.)
Cattle	0.036
Buffalo	0.036
Pig	0.078
Chicken	0.062
Human excreta	0.070

Source: Training course of IREP, organized by planning commission and Gandigram Rural Institute (2004).

2.6 Biogas Production Processes

Biogas production is a relatively slow process happening over a period of several days. The principle reaction taking place in anaerobic digester is consecutive but simultaneous. The different phases of the process are solubilization (hydrolyzing phase), acid generation (non-methanogenic phase) and methane generation (methanogenic phase) (Santra, 2001).

The first step involves the solubilization of complex organic materials constituting the digestor food stock. They are composed of carbohydrates, fats, protein, nitrogen compounds, salts and debris. In the second stage, the bacteria reduce the soluble organic material from the first step to soluble simple organic acid. In the third step methane bacteria reduce organic acid primarily acetic acid and certain other oxidized compounds to methane and carbon dioxide (Grewal et al., 2000).

2.7 Theory of Biogas Burner

Biogas can be used as a cooking fuel and in any gas- burning appliances that requires low-pressure gas (such as lamps, stoves, refrigerators etc). Biogas is a lean gas that can, in principle, be used like any other fuel gas for household and industrial purposes, the main prerequisite being the availability of specially designed biogas burners or modified consumer appliances. The heart of any gas appliance is the burner. In most cases, atmospheric- type burners operating on premixed air/gas fuel are considered preferable. This type of burner was used in this study.

A biogas burner (stove) used for cooking purposes, consists of nozzle, an air inlet, mixing chamber and fire sieve element. The nozzle is a hollow tube made of glass, metal, plastic or bamboo. As biogas passes through the nozzle, air is allowed to be drawn in to the mixing chamber. For obtaining desired flame temperature, nozzle adjustment is done by trial & error. Combustibility of gas is maximum when flame is blue (Grewal et al., 2000).

2.8 The Slurry after Digestion

The residue from biogas has been used traditionally as a soil conditioner or fertilizer because the process produces chemical forms of minerals that are more soluble than the organic forms. Residue is also used as a feed supplement (Vandana, 2004). Anaerobic digestion modifies the properties of the waste, such as: Carbon, hydrogen and oxygen are transformed in to CO_2, CH_4 and H_2O, fermentation reduces the C/N ratio increasing the fertilizing effect, the nitrogen appears mineralized and thus suitable for the plant, well - digested slurry is practically odor less and does not attract flies and anaerobic digestion deactivates pathogens and worm ova.

Compared to the source of material, digested slurry has a finer, more homogeneous structure, which makes it easier to spread. There is an increase in the ammonia content in the digested manure, which means the amount of nitrogen available for the plants is bigger in the degassed manure (Vandana, 2004).

The organic content of the digested slurry improves the soils texture, stabilizes its humic content, and intensifies its water holding capacity (Uli et al., 1989; Nielson et al., 2002; Anderson and Sorensen, 2001).

2.9 Measurement of Biogas Production

Total biogas production varies depending on the organic materials digested, the digester loading rate, the environmental conditions in the digester, design of digester, materials used for digester construction. Not all of the bio- gas energy is available for use. Energy is required to heat and mix the digester, pump the effluent, and perhaps compress the gas. Biogas production and biogas yield are the most widely used parameters to control the anaerobic fermentation process at full scale biogas plants. Moreover, pH value and the composition of the effluent biogas are also measured. Individual evaluations of the data obtained from these measurements are vital to perform a proper control of the biogas plant. The available information on biogas production, biogas composition and composition of biomass input are used for control, improve and regulate biogas plants in the future (Grewal et al., 2000).

2.10 Designing of Digesters

Designing a properly sized digester to obtain the maximum biogas production per unit of reactor volume is important in maintaining low capital construction costs. The digester should be sized to achieve desired performance goals in both winter and summer. Design goals could be maximal gas production with minimal capital investment, achieving pollution control and reduction of pathogens, or simply the production of a reasonable amount of gas with a minimum of operational attention. The uses of the slurry after the digestion process is a critical consideration since the main income to the plant can come from that material. The differences in uses also determine the digestion retention time. Criteria must be established, prior to design, since not all goals can necessarily be achieved with a single design (Hao et al., 1980; Hong et al., 1979; Umana, 1982).

2.11 Working of Fixed-Dome Biogas Plant

When the cattle dung is used as feed stock, the biogas plant is to be filled with homogeneous slurry made from a fresh dung and water in a ratio of 1:1 up to the level of the second step in the outlet chamber. As the gas generates and accumulates in the empty portion of dome of the biogas plant, it presses down the slurry of the digester and displaces it into the outlet chamber. The slurry level in the digester falls, where as in the outlet chamber, it starts rising with the formation of gas. This fall and rise continues till the level in the digester reaches the upper end of the outlet opening, and at this stage, the slurry level in the outlet chamber will be at the slurry outlet. When the gas is used, the slurry which was earlier displaced out of the digester and stored in the outlet chamber begins to return into the digester. The difference in levels of slurry in digester and the outlet chamber exerts pressure on the gas which makes it flow through the gas outlet pipe to the points of gas utilization (Grewal et al., 2000).

2.12 Selection and Layout of Pipeline and Biogas Accessories

It is often seen that the size of the pipeline for conveying the gas is not selected properly. This results in lower efficiency or higher cost to the plant owner. In designing the gas distribution system, the parameter that needs to be controlled is the gas pressure. When the gas flows in a pipe, there is a loss in its pressure due to friction. A properly designed pipeline is one which does not cause a pressure drop of more than 2-3 cm of water column under any circumstances.PVC or HDP pipe can also be used for carrying biogas instead of steel pipe. The efficiency of these pipes to carry biogas is more than that of steel pipe as it is very smooth. The cost of these pipes is also less as compared to that of steel pipe. Due to these reasons, normally HDP or PVC pipes are being used for carrying biogas.PVC pipes were used in this study due to the above reason (Biogas support programme, 1994).

According to Grewal (2000), the following points should be kept in mind while selecting and laying the pipelines:

1. The pipes and fittings to be used for laying gas distribution system must be of best quality. Additional emphasis should be given for the selection of valves to be fitted.
2. All steel pipes should be coated with protective paints to avoid corrosion.
3. As far as possible only bends (not elbows) should be used for 90^0 turns in pipelines. This reduces pressure drop.
4. Only gate valves, plug valves or ball valves should be used for the gas pipeline to minimize pressure loss during flow through the valves.
5. For connecting burners with gas pipelines, use of transparent polyethylene tubes should be avoided and as far as possible only neoprene rubber tubes should be used.
6. The over ground pipe should be laid at minimum slope of 1:100 (i.e. one cm in one meter).
7. The over ground pipes should be laid along the walls and should not be hanging free. It could

    be hooked with clamps at every 2m and should not sag at any point. A continuous slope is essential.

2.13 Transfer of the Plastic Film Biodigester Technology

More than 40 provinces in Vietnam have participated in the transfer of the plastic film biodigester system. The technology was also introduced to Cambodia, Lao, the Philippines and Thailand. More than 15,000 units have been set up in Vietnam during the past ten years;

with numbers installed annually showing a consistent increase (Duong et al., 2002).In Ethiopia construction and use of the plastic film biodigesters are yet in an infant stage(Yacob, 2000). So, more efforts are warranted to promote this technology in the country.

2.14 Promotion of Fixed and Floating Dome Biogas Plant

The national programme of India in collaboration with the khadi village industries commission in Bombay has technicians who provide on the spot training and help to get rid of bottlenecks. Constraints were also emphasized, e.g. climatic considerations, lack of accessories locally available etc. A $3m^3$ plant costs US $750, which is outside the reach of many rural people. Other major constraints are the cost of the installation, operation, and maintenance of biogas plants. Lack of technical experts and servicing personnel constitute a serious drawback. Various problems of digester construction and use of materials were described. An Indian design used steel for the digester, which of course, is very expensive, a Chinese design using earth materials has been tried, which is cheap but allows leakage of effluent, and a Tanzanian design built with oil drums has proved not to be durable(Gunnerson and Stuckey ,1986).

According to Yacob (2000), experience of biogas technology in rural Ethiopia is very limited. Thus the best solution to this problem is to promote the wide use of plastic biodigesters which are economically feasible, technically simple, technologically efficient and environmentally comfortable.

2.15 Economic Evaluations of Biogas Plants

The financial viability of biogas plants depends on whether out put in the form of gas and slurry can substitute for fuels, fertilizers or feeds which were previously purchased with money. If so the resulting cash savings can be used to repay the capital and maintenance costs and the plant has a good chance of being financially viable. However, if the out put does not generate a cash inflow, or reduce cash out flow, the plants lose financial viability (Hao et al., 1980; Hong et al., 1979; Umana, 1982).

2.16 LDPE Geomembrane Plastic

Low density polyethylene (LDPE) Geo- membrane is made of high quality LDPE and physical and chemical parameters are excellent. It is named for its high quality, flexibility and width. The product is widely used for the water proofing, such as reservoir, Channel, airport, garden and construction (Wang, 2004).

Figure 4: Geomembrane Plastic used for Biodigester Construction

Prior to liner installation all surfaces to be lined shall be smooth, free of all foreign and organic material, sharp objects or debris of any kind, the ground shall provide a firm, unyielding foundation with no sharp changes or abrupt breaks in grade. The installer, on a daily basis, shall approve the surface on which the geomembrane will be installed (Wang, 2004).

The technical parameter of the geomembrane plastic welding machine is as follows,

- Temperature Scope: O- 450 $^{\circ}$c
- Speed Scope : O.5- 4m/min
- Supply voltage: 220V consumption : 440W
- Heat- Seal strength: 85 % master material (tensile strength at cutting direction).
- Applied material: PE (LDPE, HDPE, PVC and compound geomembrane).
- Thickness of material: 0.3-1.2 mm.
    (Wang, 2004)

According to Wang (2004), the operation rule for the hot wedge welding machine is as follows
- Put the operate shaft in the Position "off" then put the plug fit in to the socket for an electrical connection.
- The over lapped width of geo- membrane should be between 80-100 mm the edge of the sheets should be cut straight and cleaned.
- Turn on the switch of speed controlling and put it in the position of "3" or so turn on the switch of temperature controlling and adjust the temperature up to 250 $^{\circ}$c or so. When the red light bright, you can put the sheets between the belts (the left sheets under the hot

wedge, the right sheets above the hot wedge), push the shaft in the position. "On", it can work now.
➢ When hot wedge welder propels at the end of the sheets, you must push the operate shaft back to the position ", to avoid damage the belt.
➢ Attention! if the surface of the pond or reserve make it propel at a uniform speed.
(Wang, 2004)

Figure 5: Geomembrane Plastic Welding Machine,(photo shoot by Author)

2.17 Theory of Environmental Impact Assessment (EIA)

Environmental impact assessment (EIA) is a process or an instrument used to forecast and consider both positive and negative environmental and social consequences of a proposed development project to their implementation (EPLAUA, 2007).

The overall goals and objectives of EIA are to promote environmentally sound and sustainable livelihood and development in the country i.e. to bring ecological, economical and social sustainability in the way of development (EPLAUA, 2007).

3 Materials and Methods

3.1. Description of the Study Area

3.1.1 Location

The research was conducted in Mersa-Chekorsa village located in Habru district, North Wollo administrative zone of the Amhara region, Ethiopia. The region has a total area of 170, 752 km^2 and a population of 14.5 million (CSA census, 1994 cited in Ayen Mulu, 2004). Mersa is located at 490 kilo meters North of Addis Ababa and 90 kilo meters North of Dessie town and the main all weather road from Addis Ababa crosses the town. Mersa is located at latitude of 11^{0}35'N, longitude of 39^{0}38'E and an elevation of 1557 m.a.s.l. The Woreda land coverage is 47210 hectare and the average altitude ranges from 1200 to 2350 m.a.s.l.

North Wollo

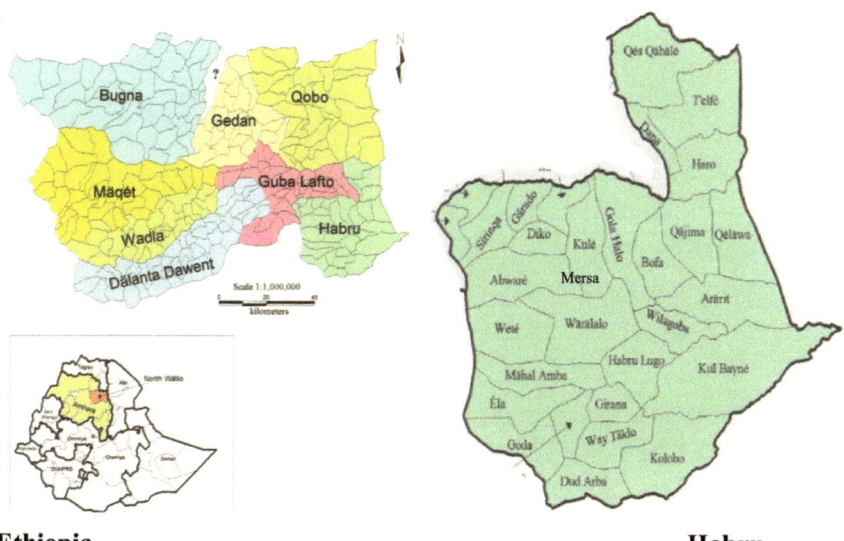

Ethiopia **Habru**

Figure 6: Map of the Study Area

3.1.2 Socio-economic activity

In 2000, the population of the Woreda was 16,209 of which 8043 were female (HARDO, 2006), and 90 % of the economic activity of the Woreda depends on mixed farming and 5.5 % on arable farming. The rest 1%, 1.5%, and 2% depend on trade, daily labour and cottage, respectively (Bech and Wavern, 2002).

3.1.3 Climate

Habru Woreda is classified under moist warm climatic zone (Bech and Wavern, 2002).The mean annual rainfall is 1090mm.The mean minimum and maximum daily temperature ranges from 13.27 ^{O}C to 28.96 ^{O}C with an average of 21.12 ^{O}C (Appendix 10). So, the temperature is mesophilic which favors the basic requirement to increase biogas production proportionately. The area has a bimodal type of rainfall. The first rainfall season is the 'Belg' (March to May) and 'Kiremt (July to September). The topography of the area is some what flat. Therefore, the area was very suitable to undertake the proposed research.

3.1.4 Land Use

The land use pattern of Habru Woreda is summarized in Table 3.1. Annual crop production, sparse vegetation and grazing land constitute about 44.9 %, 27.16% and 7.18 % of the land mass of the Woreda respectively. Only 1614 hectare is covered by forest (772 hectare natural and 772 hectare planted) and 11207 hectare is covered by shrub and bushes. The forests are owned and managed by government and cooperatives (HARDO, 2006).

Table 4: Land use in Habru Woreda

Land use type	Area coverage (ha)	Proportion (%)
Annual crop production	21194	44.89
Perennial crop production	721	1.52
Grazing land	3388	7.18
Forest land	1614	3.42
Shrub &bushes	11207	23.74
Built-up land	3173	6.72
Uncultivated (cultivable) land	1422	3.01
Waste land	4491	9.51
Total	47210	100%

Source: HARDO (2006).

3.1.5 Livestock Population

Ethiopia has the largest livestock population in Africa, totaling some 30.6m tropical livestock units (TLU).Cattle are the most important, numbering some 27m head, followed by sheep(24 m),goats (18 m),equines (7m) and camels (1m).There are also 53 million poultry in the country (Badege,2007).

According to the Ethiopian Agricultural Sample Enumeration (2002), the total cattle population in Amhara region and North Wollo Administrative Zone is 10,512,777 and 910,763 respectively.

3.2 Experimental Design and Layout

Four biodigesters were made from single & double layered cylindrical geomembrane plastic film, constructed below the ground on a trench excavated at a dimension of 7m*1.5m*0.5m and above the ground surface on a concrete block wall platform filled with stone, sand, aggregate and selected fill constructed at a dimension of 7m*1.5m*(0.75m and 0.5m) with a slope of 2^0.The main purpose of constructing the plastic biodigester below and above the ground surface is to compare the amount of gas production in both conditions and to come to the real conclusion that in which location of installation the quantity of gas production will be higher with respect to the amount of heat acquired by the biodigesters. Besides, the plastic biodigester was constructed by using single and double layer geomembrane plastic to enable to compare and proof their resistance against biogas pressure and select the best out of which. The biodigester has internal dimensions of 6.2 meter long and 1.33 m in diameter, and to provide a liquid volume in the proportion of 75% of bio digester volume. One Chinese model fixed-dome biogas plant constructed by the college was also another treatment used for comparing the geomembrane plastic biodigester. The capacities of the digesters were $3m^3$. The experiment was carried out in Mersa ATVET College, Ethiopia from Nov 20, 2006 to July, 2007.The treatments of the experiment were:

- ◊ PSA : plastic biodigester , single layered and constructed above ground.
- ◊ PDA : "   "   double   " "   "   "   "
- ◊ PSU : "   "   Single   " "   "   under ground.
- ◊ PDA : "   "   double   " "   "   "   "
- ◊ FDU : Fixed dome biodigester constructed underground.

The biodigesters were fed with cattle manure and water at equal proportions. The manure was collected from Mersa Agricultural TVET College dairy farm and the nearby private cattle shed. A constant loading rate (10.36 kg manure DM per 1 m^3 digester liquid volume) was applied to all five bio digesters (Table 5). The fresh manure was previously mixed with enough water so as to ensure a retention time of 40 days.

Table 5: Technological Parameters of the Experimental Biodigesters.

Constants	Value
Plastic width, m	2
Circumference, m	4
Internal diameter m	1.33
Loading rate, Kg ODM/m^3/day	10.36
Plastic length, m	6.2
Retention time, days	40

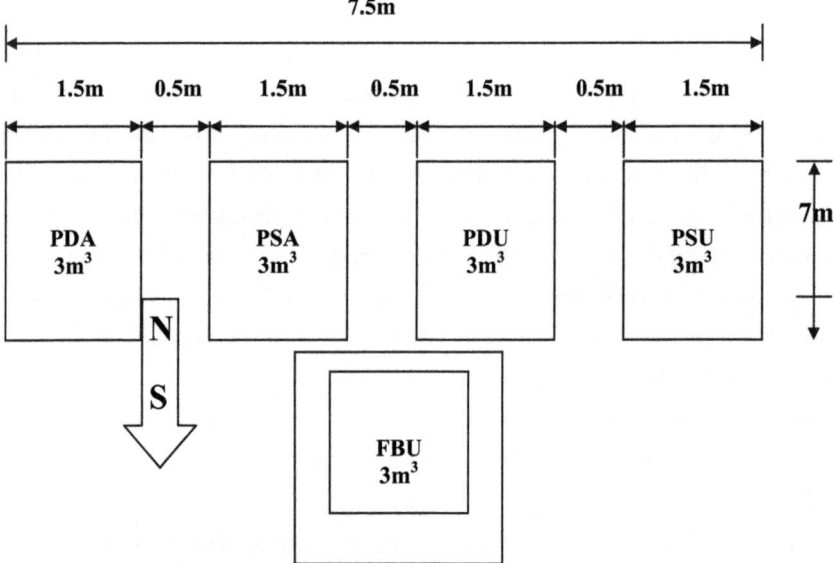

Figure 7: Layout of the Experimental Site

Figure 8: Field Installation of the Plastic Biodigester after Feeding (Photo taken by the author)

3.3 Geomembrane Plastic Construction methodology

The plastic biodigester was constructed with the use of materials like geomembrane plastic, PVC pipes, GI pipes, gate valves, reducers, GI caps, sockets, nipples, neoprene rubber hose and biogas stoves with the help of electrical geomembrane welding machine from Ambasel trading company and CM-43 adhesive with other mixtures from one technician working in North Wollo,GARDO.

Three rolls of geomembrane plastic material, was bought from North Wollo, Mersa town. Each rolls of plastic has a dimension of 13 meter width and 13.5 meter length.

The plastics were taken and placed carefully to the work shop of Mersa agricultural T.V.E.T College after the floor has been cleaned. The plastics were cut at a dimension of 7meter length and 4.50meter width. Then, it was welded with the help of electrical plastic welding machine across the length and the circular part of the cylinder was fitted with the help of CM-43 adhesive and with other chemical mixtures. Few number of small openings found in the plastic were closed with the help of vehicle inner tube and the CM-43 chemical mixtures. In this manner two double layered and two single layered cylindrical geomembrane plastic biodigester were made.

The cylindrical geomembrane plastic biodigester has a length of 6.2 meter and 1.33 meter in diameter. It has an inlet for entry of input materials, gas outlet for exit of produced gas and slurry outlet for disposal of fermented slurry. The digester and gas holder was made as one unit in a cylindrical shape from black 5 mm thick geomembrane plastic.

The digester costs EB 1056.15 and takes about 3 hours to assemble. The components and prices are shown in Table 10.For proper functioning; the digester requires the excreta from five cows dropping 15 Kg of waste on the average per day. The digester also requires an adequate water supply, ideally operating on one parts of water for one part cattle dung.

3.4 Data Collection Procedures

Different data which were pertinent to the study objectives were collected following standard procedures. The following variables were measured and analyzed during the study; amount of gas production, quantity of input and output slurry, temperature of the air, total-solids content of the fresh cow dung and digested slurry, pH of the fresh cow dung and digested slurry, total–N and organic matter content of the substrate & the slurry to determine whether the cow dung or the fermented slurry is preferable to use as organic fertilizer. Daily rainfall and temperature were recorded from meteorological station found at about 150 m distance from the experimental area. The daily temperature of the slurry was also measured using thermometer.

3.4.1. Input to the Digesters

The type of input material which was found feasible and available in the study area was cow dung as there was dairy farm at a distance of 25 meter from the experiential site. Manure inputs were measured using known volume container (i.e. bucket) of 25 liter. One bucket of cattle dung was mixed with 1 bucket of water (Grewal et al., 2000). The digester loading which indicates how much organic material per day has to be supplied to the digester or has to be digested was calculated and the average loading rate was 10.36Kg ODM/m^3/day. That is as per the design, three bucket of dung (75Kg) were mixed with three bucket of pure water (75litres) so as to produce $3m^3$ gas per day on May 5, 2007. After the slurry mixture has been fed in to the digester, 15 liters starter material prepared from cattle dung and water in 1:1 ratio and allowed to ferment for one month in a closed barrel were added at equal amount to all five biodigesters to initiate and facilitate the fermentation process (Table 6).

Table 6: Amount of Cow-Dung and Water Fed to the Biodigesters

Parameter	Plastic biodigester				Fixed- dome concrete biodigester
Biodigester type	PSA	PDA	PSU	PSA	FBU
Biodigester volume, m^3	3	3	3	3	3
Manure, kg / plant	75	75	75	75	75
Water, liter/plant	75	75	75	75	75

3.4.2 Measurement of Gas Production

The aim of measuring biogas is to determine the specific gas production obtained at specific retention times. In practice specific gas production represents the gas production of a specific feed material in a specific retention time at a specific digester temperature. The available information after measuring the amount of biogas produced from the two model biodigesters were used to compare the quantity of gas produced and efficiency of gas production in them. Thus, quantity of gas produced from the two models of biodigester were measured with the help of standard sized biogas burner which is, certified by ISO and manufactured by gas crafters in Bombe, India. The burner has a capacity of $0.45m^3$ per hour and it was adjusted with the help of its air shutter until blue flame comes to burn with its maximum capacity. Time to burn was taken by stop watch for consecutive hours of one day. So the daily gas production from the digester(s) is the sum total of the hours run by each burner and its gas consumption rate. The stove was purchased from the popularization department of the EREPDC. Water and coffee were boiled during burning & measuring time.

3.4.3 Temperature of the Air and Slurry

Temperature is one of the factors affecting the growth rate of microorganisms involved for the production of biogas. The temperature of the air was measured via mini-max thermometer in a metrological station found around the study area and the daily temperature of the slurry in the biodigesters was measured using ordinary electronic thermometer.

3.4.4 Total-Solids (DM) Content

As the water content of natural feed material varies, the dry matter content of the substrate and spent slurry was measured by drying a sample at 70 $^\circ$C in an oven and weighing the residue on a precision electronic balance.

3.4.5 The Organic Dry Matter (ODM)

Only the organic or volatile constituents of the feed material are important for the digestion process. The knowledge of organic dry matter also helps in calculating the digester loading rate which indicates how much organic material per day has to be supplied to the digester or has to be digested. For this reason, only the organic part of the dry matter content was considered and this was analyzed in the laboratory.

3.4.6 pH of the Fresh Cow Dung and Digested Slurry

The pH of the fresh cow dung and fermented manure was measured by pH-meter in the laboratory.

3.4.7 Quality of Output Slurry

The composition of digested slurry produced by anaerobic fermentation in the two models of biogas plants were determined in the laboratory. Five digested slurry samples from all five biogas plant and two fresh manure samples from the cow dung fed to the two types of biogas plant were taken by random sampling technique and the weight of samples were determined as per the laboratory standard.

According to AOAC (1990), total- N was determined by Foss- Tecator Kjeldahl procedures after taking 1 gram of manure sample and digested with sulphuric acid and salicylic acid and estimated by Kjeldahl method. The process of digestion took about three and half hours in the laboratory.

Organic matter content was determined by the use of Toffle furnace by ash method. 10 gram of well mixed manure in dry nickel crucible or silica basin was weighted and was put in a low flame or hot plate till the organic matter begins to burn. The crucible in a muffle furnace was placed at about 550 ^{O}C for 8 hours. The crucible with a grayish white ash formed was removed from the furnace cool in a desicator and weigh. Therefore, the residue represents the ash and the loss in weight represents the moisture and organic matter (Tekalign et al., 1991).

Dry matter content of the slurry and fresh cow dung was measured by oven dry at 70 ^{O}C. Fresh cow dung samples were taken on April 27, 2007 and fermented slurry samples were taken after gas was fully generated and measured i.e. on June 7, 2007 from five biodigesters and two cattle sheds.

3.4.8 The Efficiency of Biodigesters

It was evaluated by comparing its gas and slurry production with the amount of substrate fed in relation to the volume of the digester and compared by calculating their specific gas production. Specific gas production was determined by dividing the daily volume of gas measured by the amount of cow dung loaded into the plant.

3.4.9 Social Aspect of Biomass and Biogas Technologies

Open-ended interviews regarding the use of biomass (firewood, cow dung, crop residue etc.), use of other bio fuels (kerosene etc), use of alternative fuels (biogas, electricity, wind mill, solar cells, water mill etc), and level of knowledge on the operation and utilization of biogas technology in particular were carried out on 70 small and model farmers taken by stratified sampling technique in Habru district, Bohoro (administrative Kebele 9) and Mersa surrounding (administrative Kebele 10). They have low, medium and large incomes.

3.4.10 The Economic Viability of a Plastic and Fixed Dome Biogas Plant

The biodigesters out put can substitute the use of conventional fuels (Grewal et al., 2000). So, the resulting cash savings of the biodigesters which used to repay the capital and operational costs could determine the financial viability of the technology.

The slurry and gas output of the plastic and fixed-dome biodigesters were measured, valued and the gross value of output with the cost of plant construction and operation was compared to arrive at the NPV result.

Thus, the financial viability of the two models of biogas plants was computed by using cost-benefit analysis (Mazumdra, 2004).

3.4.11 The Environmental Impact of the Plastic Biogas Plant

The environmental impact of the new geomembrane plastic biogas technology was assessed and studied by computing:

1. The replacement of conventional fuels such as reduction of fuel wood consumption in relation to deforestation,
2. Green house gas emissions (i.e. CH_4 and CO_2),
3. Reduction of cow dung utilization in relation to soil fertilization and,
4. Reduction in the use of chemical fertilizers with the use of biodigester slurry as organic fertilizer.

3.5 Statistical Analysis

SPSS software was used to analyze the collected data.

There was only one replicate for each treatment due to shortage of logistics.

- Amount of daily gas production as a function of substrate input and temperature,
- Organic matter and total nitrogen content of the fresh cow dung and digested slurry,
- Specific gas production (m^3gas/m^3 substrate) as a function of biodigester type and location of installation

- ➢ pH of the digested slurry and substrate, and
- ➢ Time history of maximum and minimum ambient temperatures i.e. mean monthly and annual temperatures was tabulated and put in graphs.

4 Result and Discussion

4.1 Operation of Plastic Biodigester

Consecutive but simultaneous phases of anaerobic digestion of the homogenous dung mixture by various types of anaerobic microorganisms' have taken place for 40 days depending on the prevailing local average temperature of 21.4 ^{O}C. After the fermentation process has completed, gases and nutrient rich sludge were accumulated and contained in the upper one third, and in the lower two third portion of the biodigester bag respectively (Figure 9).

The out put has provided 6- 7 hours of cooking time of a single Indian burner of 450 liters per hour capacity.

Figure 9: Plastic Biodigester at the Beginning of Gas Generation (Photo taken by the author)

4.2 Biogas production

Gas was burnt and measured after gas has completely produced within the designed HRT of 40 days with the help of calibrated biogas burner and stop watch (Figure 10).As can be seen in Figure 11, gas production as the proportion of biodigester liquid volume was higher for a single layered and above ground plastic biodigester than others and very less amount of gas was measured from the fixed-dome biodigester.This was because more sun light temperature (27.65 ^{O}C-32.7 ^{O}C) was absorbed in a black geomembrane plastic sheeting digester than reinforced concrete fixed-dome biodigester (Table 7). Temperature is very important factor which positively or negatively affects the activity of microorganisms in the production of biogas. Thus, geomembrane plastic biodigesters have generally produced higher amount of gas than fixed-dome biodigester.Moreover, the construction and use of single layered above ground geomembrane plastic biodigester is preferable than other geomembrane plastic biodigester.

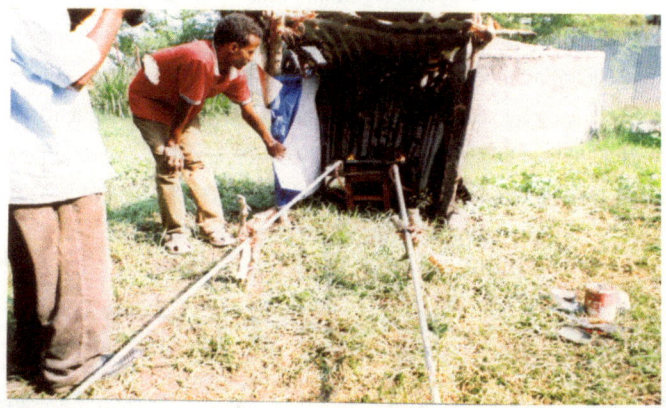

Figure 10: Burning and Measuring of Biogas with a Biogas Burner after Gas Generation (Photo taken by the author)

Table 7: Total Values for Biogas Production (9 Am to 4 Pm) in Bio-digesters with different Types, Layers and Location of Installation.

BIOGAS PRODUCTION	BIODIGESTER TYPE					
	PDA	PSA	PDU	PSU	FBU	Average for plastic digesters
Hours recorded during burning of gas in a burner /plant	6.33	6.75	5.75	5.50	5.25	6.08
Liters/day/plant	2850	3037.5	2587.5	2475	2362.5	2737.5
Specific gas production per day (m^3/kg).	0.0380	0.0405	0.0345	0.033	0.0315	0.0365
Average daily slurry temperature, ^{O}C.	29.99	32.7	29.18	27.65	25.96	29.88

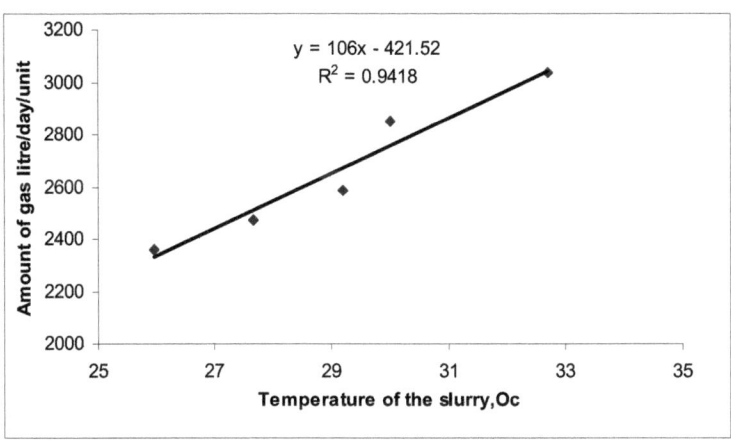

Figure 11: The Relationship between Temperature of the Slurry and amount of Gas Production for the Plastic and Fixed-Dome Biodigesters.

Table 8: Comparison of Average Slurry Temperature, OC and amount of Gas Produced, m3 / Day of the Biodigesters

Types of GPBD	Average Slurry temperature,^{O}C of GPBD	Amount of gas produce, m^3/day	GPBD,^{O}C– FBU (25.963 ^{O}C)	GPBD,m^3/day – FBU,2.36m^3/day
PSA	32.700	3.04	6.737	0.68
PDA	29.988	2.85	4.025	0.49
PDU	29.175	2.59	3.212	0.23
PSU	27.650	2.48	1.687	0.12

4.3 Temperature of the Air and Slurry

The average minimum and maximum atmospheric temperature during fermentation time between May 5, 2007 and June 14, 2007 was 15.75 ^{O}C and 33.34 ^{O}C respectively(Appendix 14), which is in the range of mesophilic temperature, and according to Grewal et al. (2000), temperature is one of the factor affecting the growth rate of micro-organisms involved in the production of biogas and effective and efficient anaerobic fermentation is carried out in mesophilic temperatures averaging between 24 ^{O}C and 45°C.

The local average atmospheric temperature of Mersa was 21.12 ^{O}C according to 12 years of data (Appendix 12) which was very suitable for biogas production.

According to Hu Qichun (1991), a biogas plant could perform satisfactorily only where mean annual temperatures are around 20 ^{O}C or above or when the average daily temperature is at least 18 °C. With the range of 20-28 °C mean temperature, gas production increases over proportionally. If the temperature of the biomass is below 15 °C, gas production will be so slow that the biogas plant is no longer economically feasible. So, the temperature recorded in the study area was very suitable for normal fermentation and higher amount of gas was produced by mesophilic bacteria (Table 8).

The average temperature of the fermented material in the biodigesters were in the range between 25.96 °C and 32.7 °C as described in (Figure 12), which is greater than the critical 15 °C.Thus, gas was completely produced within the designed HRT value of 40 days. As it is mentioned in Figure 12,

R^2 is 0.945. Thus, 94.5 % of the variability in the amount of gas produced by the biodigesters was due to the variation in the slurry temperature.

The average temperature of the slurry measured in the geomembrane plastic biodigesters exceed by 1.687 °C to 6.737 °C the slurry in fixed-dome biodigester (Table 8). Thus, geomembrane bag was observed to have the best advantage of heating the digester contents easily and produce higher amount of gas (0.12-0.68 m^3 of gas per m^3 of digester per day) than fixed-dome biodigester made from reinforced concrete (Table 8). Since its walls are thin and black in color, it can be heated quickly with an external heat source, such as the sun radiation of the same degree in the study area. Similar results were reported by Bui Xuan and Preston (1995); that found average temperatures in bag digesters, compared with dome types, are 2 °C - 7 °C higher in the bag (0.235-0.61 m^3 of gas per m^3 of digester per day).

4.4 Characteristics of Bio-digested Slurry (Effluent) and the Influent

Organic substances passed through biogas plants not only produce a source of energy, but also a large quantity of digested slurry, which provides excellent organic fertilizer. The net weight of slurry discharged daily from the geomembrane plastic and fixed-dome biodigester were 123 Kg and 112.5 Kg respectively. Thus, the annual slurry output was 44895 Kg (45 tones i.e. 18% of the total substrate was lost) and 41062.5Kg (41 tones i.e. 25% of the total substrate was lost), which was produced from 27375 kg of fresh dung fed to the digester annually, which in turn is equivalent to 13468.5 kg of dried dung cakes according to Appendix 3.Likewise, the loss in the amount of slurry have been reported by UNESCO (1982), and states that during digestion, about

20% of the total slurry is volatilized. Thus, geomembrane plastic biogas plant produced higher amount of slurry than fixed-dome biogas plant of the same capacity and using equal amount of input material.

So, with the use of geomembrane plastic biodigester, the farmers would get 45 tones of fermented slurry which could be used to apply on the farmland as organic fertilizer. The slurry could increase the soil's nutrient retention capacity (or cation exchange capacity), improves the physical condition by increasing the water holding capacity, and improves the structure. Not only does fermented slurry replenish soil fertility, but it also helps to maintain or create a better climate for soil micro-flora and fauna. It is the only avenue available to farmers for improving the soil's organic matter.

Thus, by conversion of cow-dung in to a more convenient and high-value fertilizer (biogas slurry), organic matter is readily available for agricultural purposes, thus protecting soils from depletion and erosion.

Recent research has shown that the effluent from biodigesters is a better fertilizer than the original manure when applied to crops such as cassava and duckweed (Le Ha Chau, 1998a,b) or when used in fish ponds (Pich and Preston, 2001).

Therefore, the farmers should be advised to use geomembrane plastic biogas plant so as to utilize the dung for dual purposes such as the produced gas as fuel for cooking and the remaining large quantity of slurry as organic fertilizer which helps to increase crop production by preventing it from burning in the form of dry dung cake and ashing down the good manure creating unhygienic conditions in the kitchen.

4.5 Characteristics of Total-N in the Slurry and Influent

The average total nitrogen of the substrate and digested manure of the geomembrane plastic and fixed-dome biodigester were 0.37%, 1.13% and 1.15% consecutively as described in Table 9. Similarly, Grewal et al. (2000), reported that the total-N content of fresh dung as 0.242% which is 34.6% less than the result of this study.

Thus, fermented slurry contained larger nitrogen content than fresh cow dung in both models of the biodigesters because at the beginning of fermentation in biogas plants by anaerobic bacteria, the number of nitrifies were few but increased through time and reached maximum after some days of fermentation. At this stage, the nitrogen deposited in the cow-dung mixture was consumed by those micro-organisms for their reproduction and metabolism. Thus, the amount of ammonium in the manure has decreased. As fermentation continued and increased, the number of nitrifies have decreased and died at last. Thus, the amount of ammonium in the slurry increased due to the completion of fermentation and the death of micro-organisms. Moreover, in an air tight biogas digester more organic acid such as acetic acid, prop ionic acid, butyric acid, ethanol and acetone was produced doting anaerobic fermentation of soluble simple organic substances which helps to absorb and fix ammonia and minimize the lose of nitrogen thus conserving the fertility of the manure. So, the higher quantity of nitrogen was converted in to the useful nitrate and ammonia which is the most important ingredient for plant growth. There was a greater conversion of organic substrate to nitrogen during 40 days of the experiment because microbial anaerobic degradation was facilitated with higher temperature.

According to Hu Qichun (1991), during anaerobic fermentation, part of the total nitrogen is mineralized to ammonium and nitrate. Thus, it can be more rapidly taken up by many plants and in a number of applications, slurry from biogas plants is even superior to fresh dung especially when the slurry is spread directly on fields with a permanently high nitrogen demand (e.g. fodder grasses) or when using slurry compost to improve the structure of the soil.

Fermented slurry has larger nitrogen content and hence, high fertilizer value in increasing the fertility of the soil than fresh cow dung. Therefore, it is recommended to use fresh dung as an input material for biogas plant than using it for compost making and as dry cow dung cake. According to Grewal et al. (2000), in case of air-drying of animal wastes, 30-35 percent of nitrogen is lost in air whereas through anaerobic fermentation route, it is limited to 11 to 15 percent. Processing evidence suggests, however, that slurry is much more effective than dung when applied as fertilizer, French(1979),discusses that slurry is 13 % more effective than dung, and Van Buren (1974), reported that the ammonia content of organic fertilizer fermented for 30 days in a pit in china increased by 19.3 %.

As can be seen in Figure 13, the single layered geomembrane plastic biodigester constructed above ground produced higher amount of total nitrogen than others. This was due to higher amount of slurry temperature (Table 8) absorbed in the single layered above ground biodigester activated anaerobic micro-organisms to convert more simple organic substances of the substrate in to simple organic acid so as to fix more ammonia. Therefore, total-N was affected by material and position of biodigester construction. Therefore, it is recommended to construct a single layered above ground geomembrane plastic biodigester than the Chinese model fixed-dome biogas plant.

Mersa ATVET dairy farm fermented cow manure contains 1.0164%- 1.2026% N and 50-75% Organic matter. Similarly, DebreZeit cow manure contains approximately 1.5% N and 1.3 % P per unit dry weight (Newcombe, 1983). Thus, the annual output of slurry equivalent to 13468.5kg of dry dung cake converted to 152.2 kg of N and 8127 kg of organic matter on the average.

In the slurry which has higher total nitrogen content there is higher proportion of ammonium which constitutes the more valuable form of nitrogen for plant nutrition (Werner Kossmann et al.,2007).

Certainly, the N-content of the digested slurry was regarded as comparable to that of chemical fertilizers (Urea 48 % N and DAP 46% N, 18% P). Thus, fermented slurry is very relevant as an input material to sustainable agriculture.

Table 9: Effect of Material & Position of Biodigester Construction on the Composition of the Effluent.

Components of the effluent and influent	Biodigester types and fresh cow dung fed to the biodigesters							
	PSU	PDU	FBU	PSA	PDA	IFB	IPB	Average For Plastic digesters
Organic matter,%	50	50	66.7	75	60	80	85.7	58.75
Organic matter,Kg	37.5	37.5	50.025	56.25	45	60	64.28	44.06
Organic carbon,%	29	29	38.686	43.5	34.8	46.4	49.706	34.08
Total N,%	1.148	1.1368	1.1536	1.2026	1.0164	0.3626	0.3822	1.126
Total N, Mg/kg	11480	11368	11536	12026	10164	3626	3822	11259.5
Dry matter,%	0.2	0.2	0.5	0.6	0.4	0.7	1.0	0.35
Ph	6.8	6.9	7.2	7.1	7.6	6.8	7.0	7.1

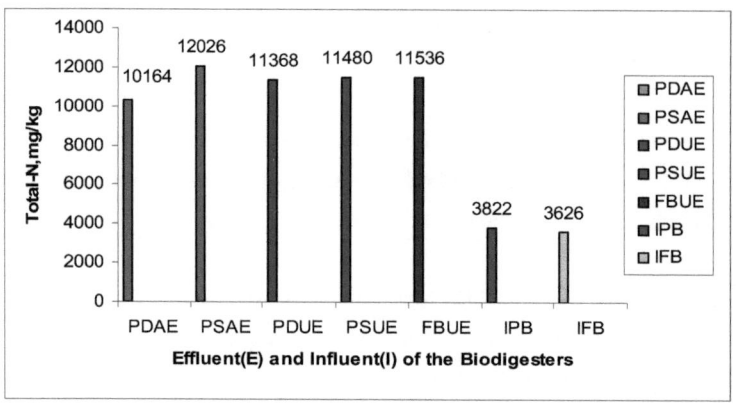

Figure 12: Comparison of Total-N Fresh Cow Dung & Fermented Slurry for Plastic and Fixed Dome Bio-digesters.

4.6 Characteristics of Organic Matter in the Slurry and Substrate

The application of digested slurry to crop serves as a dual purpose; a soil conditioner as well as a source of plant nutrients. The digested slurry, besides furnishing plant food, is beneficial to the soil as it increases the water holding capacity and improves its structure. The solid excreta of dairy cattle are used for making farmyard manure and for fuel purpose, but the urine is invariably not utilized due to its liquid nature and difficulty in handling. According to Figure 14, the PSAE showed larger organic matter content than FBUE. This could be because the amount of organic matter in the influent of plastic biodigester (IPB) was higher than that in the influent of fixed-dome biodigester (IFB).The organic matter and dry matter content of the fresh cow dung and fermented slurry as per the analysis were 80%-85.7%, 0.7%-1% and 50%-75%, 0.2%-0.6% as described in Table 9. Even if higher value was recorded for fresh manure, the use of excreta of dairy cattle in biogas plants was found to be better than the farmyard manure in several ways. A part of nitrogen which is ammonia, found in the slurry becomes available to the plants. The ability of the wet digested slurry to aggregate soil particles immediately after application is also very unique. The digested manure was available in 40 days as compared to 4-6 months taken in the usual method of composting in a manure pit.

According to Grewal et al. (2000), almost all plant nutrients are retained in the digested slurry in such finely divided state that, it mixes up with the soil quickly and thoroughly, and the soil bacterial activity increases substantially. Thus the application of digested slurry gives better yields for all crops as compared to farmyard manure (FYM) made from the same quantity of cattle dung. Therefore, for soils which have low organic matter content, the application of organic manures such as anaerobic digested slurry is highly recommended.

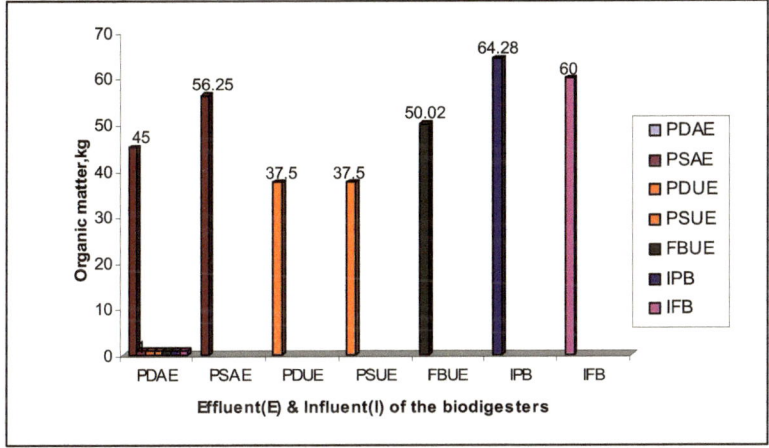

Figure 13: Comparison of Organic Matter (Kg) Content of Fresh Cow Dung and Fermented Slurry for Plastic and Fixed Dome Biodigesters.

The digested slurry in this study was thin and used directly to the crops through the irrigation channels and by direct splashing on the farmland. Moreover, it was also put in a fish pond of Mersa agricultural T.V.E.T. College and fishes were nourished. Thus, it was proved that, the waste that comes out of the digestion process as slurry was very useful both as feed and organic fertilizer. Studies by Sokoine Agricultural University in Tanzania have shown that slurry from biogas improves productivity of land and maintains soil quality that can support crop production over a long period of time (SURUDE, 2002).

4.7 Characteristics of pH of Fermented Slurry

It is generally recommended that the pH inside the biodigester should be above seven for normal fermentation and maximum gas production (Grewal et al., 2000). In the present study, the pH of the fermented slurry and fresh cow-dung was 6.8 - 7.6 and (Table 9) with a hydraulic retention time of 40 days. Thus, the condition inside the digester was very comfortable for anaerobic micro-organisms to accomplish normal fermentation and higher gas production. Since, the fermented effluent has a proper range of pH value thus making the manure more fertile.

Due to the buffer effect of carbondioxide-bicarboate (CO_2-HCO_3) and ammonia-ammonium (NH_3-NH_4^+) the pH level is rarely taken as a measure of substrate acids and/or potential biogas yield (Werner Kossmann et al., 2007).

Safely et al. (1987), used poultry manure in a biodigester with a hydraulic retention time of 22 days at 35 ^{O}c, and observed an average pH of 7.2. Similarly, Fischer et al. (1979), studied methane production from swine manure in a pilot- size digester, and obtained pH values in the effluent of 6.5 to 7.6 at 35 ^{O}c with a hydraulic retention time of 15 days. Moreover Buixuan et al. (1997) reported a pH from 6.8 to 7.5, with biodigesters of an average length of 10.2 m and a loading rate of 0.7 kg DM manure/m^3.

4.8 Efficiency of the Biodigesters

The average specific gas production of the single layered above ground geomembrane plastic biodigester was greater than others (Table 7). That was due to higher amount of heat was absorbed by the digester (32.7 ^{O}c) that used to facilitate the activity of micro-organisms and increase the amount of gas production per weight of the substrate (Figure 12).

The average specific gas production obtained from the $3m^3$ sized geomembrane plastic biodigester was also greater than fixed-dome biodigester made from reinforced concrete (Table 7).Because comparatively low amount of temperature (25.96 ^{O}c) was absorbed (Table 7). Thus, the construction of single layered aboveground biodigester is better than other biodigesters as efficiency of conversion of substrate organic matter to biogas was higher.

4.9 Economic Evaluations

The design output of methane gas from the plastic and concrete biodigester was $3m^3$ day^{-1}. In order to compute its value in terms of the value of traditional fuels saved by utilizing biogas, it was necessary to estimate the amount of dried dung cakes or fire wood needed to produce an equivalent amount of energy. Thus, according to UNISCO (1982); Van Buren (1974), used in this study assume that $1m^3$ biogas substitutes for 3.47 kg of fire wood, 12.3 kg of dry dung fuel and 0.62 litre of kerosene oil. Moreover, economic evaluation was done for the five alternative biodigesters using method which is the most commonly employed method used by many extension officers. It was able to evaluate the relative advantage of a plastic biogas plant investment as compared to fixed dome biodigester on the basis of the anticipated minimum interest rate and economic life for the alternative designs.

A discount rate of 18 % has been applied throughout the analysis according to Amhara regional state loans and saving enterprise (2007). As the research life time has been limited to 1 year, however, it was not deemed necessary to look for a most precise discount rate.

Table 10: Summary of Market Value of Inputs and Outputs Used in the Analysis.

Inputs	Market value (price in EB)
Dung (EB/kg)	0.50
Water (EB/5 m^3)	2.00
Labor (EB/day)	15.00
Outputs	
Biogas (EB/litre)	3.47
Slurry (EB/tone)	90.62

4.9.1 Market Price of Inputs

Dung: According to Senait Seyoum (2007), the cost of dung was estimated in terms of

◊ Dung's value as fertilizer determined by the cost of an equivalent amount of commercial fertilizer, or
◊ The market value of dung cakes, if dung is sold.
◊ 1 ton of DAP is 16 tone of dry manure.

Price of 1 tone of DAP according to 1999 E.C was 1450 ET Birr. This was determined from a receipt voucher issued to Mersa agricultural T.V.E.T College purchasing office.

The price of dried cow dung according to Woldya market in 2007 ranged from EB5.00 to EB8.00 per 50kg sack, averaging EB 0.50 per kg of dried dung.

Water

This was valued according to the price charged by the water authority of Habru District, Mersa town. I.E. EB 2.00 per 5m^3.

Labor

The Labor used to collect water and spread slurry was valued at EB 15 day^{-1}.

4.9.2 Market price of outputs

Biogas

The biogas produced by the digestor was valued at the market value of fire wood or dried cow dung cakes, which it replaces. The observed price of fire wood in Mersa was between EB 15 and EB 20 per bundle weighing 15-20 kg. It was estimated that firewood averages EB1.00kg^{-1}. According to UNESCO (1982); Van Buren (1974) states that, the substitution rates assumed that 1m^3 of biogas substitutes for 3.47 kg of firewood or 12.3 kg of dry dung fuel. From this it was concluded that the price of 1m³ of biogas which gives the same amount of energy as 3.47kg of fire wood is EB 3.47.

Slurry

Output digested slurry was valued at the official price of DAP. N and P contents of diammonium phosphate (DAP) roughly approximated the proportions of these nutrients in dried cow dung. According to Senait Seyoum (2007), 1tone of DAP is roughly equivalent to 16 tone of dry manure. The 1999E.C price of DAP was EB 1450 tone^{-1}, thus each tone of dry cow dung which has less nutrient content than the fermented slurry is worth EB 90.62.Therefore, the cost of 1 ton of fermented slurry was assumed to be EB 90.62 to the minimum.

Three important technical assumptions were made with respect to gas production and use, and the efficient use of inputs and their conversion in the analysis.

1. The average daily gas production for a year from the geomembrane plastic and fixed-dome biodigesters was assumed to be 2.74 m³ and 2.36m³ as to the measurement taken once in drier months, but this could vary considerably with daily ambient temperature fluctuations in a year.
2. In the analysis it was assumed that all gas produced would be used, and it would have the same use as the dry dung or wood displaced by biogas.
3. The amount of slurry produced daily from the geomembrane plastic and fixed-dome biodigesters was 123 tones and 112.5 tones.

As per the design, the total quantity of fresh dung required for 3m³ size biogas plants in one year was 27375 kg or 27.375 tones but the quantity of fermented slurry collected after digestion and gas measurement was 45,000 kg and 41,000 kg in the geomembrane plastic and fixed-dome biodigesters respectively. Thus, approximately 10 tones and 14 tones of digested slurry (18% and 25%of the input mixture) were lost during fermentation from the total of 55 tones of slurry mixture available in the geomembrane plastic and fixed-dome digesters consecutively. The loss in the weight of the slurry was due to the loss of solids during fermentation.

Similarly, UNESCO (1982), states that, during digestion, about 20% of the total slurry is volatilized. Thus, about 80% of the manure is collected from fresh dung. According to Grewal (2000), the loss of solids in the biogas plant rarely exceeds 27 percent even when maximum gas is generated.

4.9.3 Cost-Benefit Analysis of Biogas Plants

Table 11: Initial Cost of Investment for Geomembrane Plastic Biodigester in EB.

NO.	Cost Description	PSA	PDA	PSU	PDU
1. Cost of Civil Construction					
1.1	Inlet and Outlet Accessories	114.00	114.00	114.00	114.00
1.2	Materials for platform construction	200.00	200.00	200.00	200.00
1.3	Labor for plant construction & installation	60.00	120	60	120
1.5	Labor for pit excavation	-	-	45	45
Sub Total 1		**404**	**464**	**419**	**479**

	2. Digester and Gas holder				
2	Geomembrane Plastic	421.60	843.20	421.60	843.20
3	Welding	10.00	20.00	10.00	20.00
4	Adhesive Chemical	210.00	280.00	210	280
Sub Total 2		**641.00**	**1143.20**	**641.00**	**1143.20**

3. Gas pipelines and appliances					
5.	Gas Pipeline & Appliances	310.55	310.55	310.55	310.55
Total Initial Investment Cost		1356.15	1917.75	1370.55	1932.75
Annual Discounted Investment Cost At Discount rate of 0.18 for 10 years service life in EB.		301.7632	426.728	304.9674	430.066

Table 12: Operating Cost for Geomembrane Plastic Biogas Plant in EB per year.

S. No.	Annual working Costs (Birr)	Use of dung as							
		Manure (Birr)				Fuel (Birr)			
		PSA	PDA	PSU	PDU	PSA	PDA	PSU	PDU
i)	digester & gas holder (10 year life) Depreciation 5%	64.1	114.32	64.1	114.32	64.1	114.32	64.1	114.32
ii)	Gas pipeline and appliances (20year life) depreciation 5%)	15.53	15.53	15.53	15.53	15.53	15.53	15.53	15.53
iii)	Civil construction cost (10 year life) Depreciation 2.5%	40.4	46.4	41.9	47.0	40.4	46.4	41.9	47.0
iv)	Maintenance (1% of total	13.56	19.08	13.71	19.23	13.56	19.08	13.71	19.23

	cost)								
v)	Cost of dung as manure, 30 tone	4077.9	4077.9	4077.9	4077.9	-	-	-	-
vi)	Cost of dung as fuel	-	-	-	-	3470.35	3470.35	3470.35	3470.35
vii)	Cost of labor	522.2	522.2	522.2	522.2	522.2	522.2	522.2	522.2
	Total Operational Cost	4733.69	4795.52	5257.53	4799.18	4126.14	4177.97	4127.78	4191.63

Table 13: Summary of total annual discounted costs for geomembrane plants in EB per year.

NO.	Costs	PSA	PDA	PSU	PDU	PSA	PDA	PSU	PDU
1.	Annual Initial Investment Cost	301.76	426.72	304.96	430.06	301.76	426.72	304.96	430.06
2.	Total operational cost for use of dung as manure	4733.69	4795.52	5257.53	4799.2	-	-	-	-
3.	Total operational cost for use of dung as fuel	-	-	-	-	4126.14	4177.9	4127.78	4191.6
	Total Cost	5035.5	5222.3	5562.5	5229.2	4427.90	4604.7	4432.74	4621.7

Table 14: Benefit Obtained from the Geomembrane Plastic Biodigesters in EB per year.

NO.	Benefit	PSA	PDA	PSU	PDU
1.	Biogas	3847.15	3609.67	3134.71	3277.20
2.	Slurry	4077.9	4077.9	4077.9	4077.9
	Total Benefit	**7925.05**	**7687.57**	**7212.61**	**7355.1**

Table 15: Initial Cost Investment for Fixed-Dome Biodigester

No.	Cost Description	Cost in EB.
i	Cost of civil construction	7689.45
ii	Cost of gas pipeline and appliances	310.55
iii.	Cost of labor for excavation and construction	900
	Total Initial Investment Cost	8900
	Annual Discounted Initial Investment Cost at Discount rate of 0.18 for 20 years of service	1662.698

Table 16: Operating Costs for Fixed-Dome Biogas Plant in EB per year.

S. No.	Annual working Costs (Birr)	Use of dung as	
		Manure (EB)	Fuel (EB)
i)	Civil construction cost (20 year life) Depreciation 2.5%	384.47	384.47
ii)	Gas pipeline and appliances (20year life) depreciation 5%)	15.53	15.53
iv)	Maintenance (1% of total cost)	89	89
v)	Cost of dung as manure	3715.42	-
vi)	Cost of dung as fuel	-	2989.06
vii)	Cost of labor for feeding dung mixture and transporting slurry	522.19	522.19
	Total Operational Cost	**4726.62**	**4000.26**

Table 17: Summary of total discounted cost for fixed-dome biogas plant in EB per year.

NO.	Discounted Costs	Fixed-Dome Biodigester	
		Dung as manure	Dung as fuel
1	Annual Initial Investment Cost	**1662.698**	**1662.698**
2	Total operational cost	4726.62	4000.26
	Total cost	**6389.32**	**5662.96**

Table 18: Benefit Obtained from Fixed-Dome Biogas Plant

NO.	Benefits	Fixed-Dome Biodigester
1	Biogas	2989.06
2	Slurry	3721.08
	Total Benefit	**6710.14**

Table 19: Summary of total costs and total benefits of the two model biogas plants

NO.	Biogas Models	Total Benefit	When used as fuel wood		When used as manure	
			Total Cost	Net Benefit	Total Cost	Net Benefit
1	PSA	7925.05	4427.90	3498.05	5035.45	2889.6
2	PDA	7687.57	4604.69	3082.88	5222.25	2465.32
3	PSU	7212.61	4127.78	3084.83	5562.47	1650.14
4	PDU	7355.1	4191.63	3164.1	5229.25	2125.85
5	FBU	6710.14	5662.96	1047.18	6389.32	320.82

In general the net benefit of geomembrane plastic biogas plants is greater than fixed-dome biogas plant and in particular the net benefit gained from PSA is greater than others. Thus, the use of geomembrane plastic biodigesters is profitable than fixed-dome biodigester.

4.10 Social aspect of biogas technology

4.10.1 Income generation through increased crop production

The waste that came out of the digestion process as slurry was rich in organic compounds. According to Table 4.5, fermented slurry has 50%-75% and 66.7% organic matter and 1.016%-1.203% and 1.154% total-N from plastic and fixed-dome biodigesters respectively. According to Grewal (2000), the organic matter in the manure can activate the microorganism in the soil. About 80-90% of the nitrogen exists in organic form. The organic matter in the slurry can be transformed in to humic acid which improves soil granular structure. The improvement of the arability in relation to moisture, fertility, air and heat is effected by the formation of stable granular structure of soil which results from the combination of organic colloid with inorganic colloid. The content of organic matter in soil is an important measure of soil fertility. From this we can see the importance of applying organic manure in the improvement of soil structure and the increase of soil fertility. Thus producing crops, vegetables and fruits with the use of fermented slurry as fertilizer could boost the production and productivity of the produce. Similarly, studies by Sokoine Agricultural University in Tanzania have shown that slurry from biogas improved productivity of land and maintains soil quality that can support crop production over along period of time (SURUDE, 2002). Therefore, popularization of geomembrane plastic biogas plant technology in the rural areas of Ethiopia could help the farmers in getting benefit and then recognizing the advantages of biogas fertilizer owing to the improvement of soil fertility and increasing crop yield.

4.10.2 Income Generation through Cost Saving

The use of biogas plant could reduce the amount of money expend for buying firewood and cow dung cake. According to this study, a household will save a minimum of EB 910 that would have been expended to purchasing firewood and cow dung cake for lighting and cooking to household use. Moreover, the household will also get enough organic fertilizer (41-45 tones) to be applied on the farm land so as to increase the fertility of the soil and thus crop production. This indicated that biogas stoves not only reduce family expenditure but also it has a significant contribution for improving the household economy through increasing crop yield. This in turn improves the livelihood of the people.

4.10.3 Perceptions of Habru Woreda People regarding the use of Biomass & Biogas Technology

According to ESP (2001), in North Wollo, Habru Woreda, the percentage of houses using mainly firewood and leaves and dung (manure) as fuel for cooking are 95 and 5 respectively. Based on the interviews and discussions carried out on 70 small and model farmers of the Habru district, in Bohoro (number 9 administrative Kebele) and Mersa surrounding (number 10 administrative Kebele) in July,2007, fuels used for cooking were wood,charcoal,leaves,maize and sorghum stack, straw and cattle dung. They used biomass fuels beyond their regenerative limits (unsustainable manner). They also used fossil fuels such as kerosene only for lighting purpose with the use of 'Kuraz' and 'Fanos'.Apart from being not available and affordable, the problem of kerosene is that it cannot be used for making *injera*.Electricity which could be used for lighting and cooking purpose is completely lacking in Bohoro Kebele but there are some people who used electricity around Mersa town. They did not use alternative fuel sources such as biogas, wind mill, water mill and solar energy technologies. Besides they did not have any idea and knowledge regarding those alternative fuel sources. They hardly used chemical fertilizers for their farmland as it is expensive that they could not afford to buy and use.

Mersa A.T.V.E.T college has invited individuals from Habru Woreda Administrative Office, Habru Woreda Agricultural and Rural Development Office, North Wollo Zone and Habru Woreda Women Affairs Office, Mersa town municipality office,10 model farmers of Habru Woreda and Mersa A.T.V.E.T College staff for the farmers field day arranged on June,2007.On the occasion, I showed and explained briefly the new geomembrane plastic biodigester and the fixed-dome biodigester technologies benefit, construction ,operation and manintainance.Thus,from the discussions with invited model farmers and other guests, I noted that, they used biomass fuels in unsustainable manner which was the cause of deforestation, land degradation and environmental pollution. Thus, it would be a mandatory to promote efficient burning devices and alternative energy technologies such as biogas plant. Moreover; they showed interest to use the new geomembrane plastic biogas technology.

4.11 Technological aspect of geomembrane plastic biodigester

4.11.1 Sustainability

According to this study the low cost cylindrical geomembrane plastic biodigester systems are cheaper and affordable by poor farmers. The material cost including installation was US$ 115. There was hardly little cost on operations and maintenance.

A survey was made of 2000 households having installed polyethylene plastic film biodigesters since 1995 in Vetnam.Most of the units were still working up to five years after installation and some even longer, where there had been good protection (Duong Nguyen Khang and Iming Tuan, 2002).Accordingly, a stronger geomembrane plastic biodigester could stay more than five years with safety operation and usage.

4.11.2 Simple technology

The geomembrane plastic film biodigester is simple and easy to construct from readily available materials and its operation does not require special skills. It was constructed by materials that are cheap and obtained easily in the market. Farmers can do the set up by them selves. It can be fixed by one or two people in about 2-3 hours. It can be easily transported from the place where it is manufactured to the interior rural regions and uncomplicated cleaning, emptying and maintenance were the good feature of the technology. This means the technology used is easily localized and institutionalized at household level.

4.11.3 Demand driven

Biodigester technology enhances energy supply decentralization, for instance rural communities having difficulties or no access to commercial fuels can produce biogas to meet their fuel needs, thereby increasing reliability of fuel supply and enhancing energy security to the rural areas. In Ethiopia greater than 90% of the people use biomass fuel since they do not have access to electricity and they could not afford to buy fossil fuels as they are expensive. Moreover this day as the supply and demand of firewood is not balanced; there is an increase in the price of firewood increases. Thus it is encouraging to promote the low-cost geomembrane plastic biogas plant which could help the farming community in bringing reliable and cleaner renewable energy sources for cooking and lighting purposes.

It was proved that, there would be an individual household demand for biogas units when awareness raising extension activities on the technology is done widely and effected well according to the interviews made in this study on 70 farmers of Bohoro administrative Kebele and Mersa surrounding.

4.11.4 Replicability

The plastic biogas technology comes from Vietnam with some major modifications. It has since been replicated in Tanzania, Uganda, Cameroon and Nigeria; projects that are replicable have sustainability potential.

The average dimensions of the geomembrane plastic biodigester used in this study were 6.2meter and 1.33meter in diameter with the measured value of 2.66m^{3}gas per day. The investment cost is around US$115, which is simple for an individual household to invest on the technology; its return period is 3.57 months on the average that the expenditures on the technology could be paid back within a very short period of time. Comparatively, the total amount of capital invested in an Indian model of fixed-dome biogas plant were EB 8000 which is very expensive and unaffordable by our farmers; and its return period is 3.17 years on the average. This means the investment cost on this technology could be recovered after 3.17 years of installation and use of the technology. This is comparatively long period of time than the case of plastic biodigester.Thus, investment on fixed-dome biogas plant would be difficult at household level but investing on plastic biodigester is encouraging for the farming community. Benjamin Jargstorf (2004) reported that a fixed-dome biogas digester constructed in Ethiopia would cost approximately EB 7000 which is very expensive. According to Thong (1989), cited in Livestock research for Rural Development (1997), the most important problem in biogas programs in developing countries has been the price of digester plants. For example the price of a concrete digester plant installed for an average family in Vietnam varied from 180 to340 US$. The size of investment is considered unaffordable by average farm families (Buixuan et al., 1997). Chinese designers tried to reduce the cost of red- mud digesters to 25-30 US$/ m3 (Gunperson and stocky, 1986), but it was still high in comparison with the polyethylene digesters (5 US$ /m^3). This is obviously one important feature, which makes the polyethylene digesters attractive. Similarly, the geomembrane plastic biogas plant would be very attractive as it is cheap and the expense of the digester plant could be paid back within slightly less than 3.57 months.

4.12 Technical problems with the geomembrane plastic digester

Main causes of damage to the digesters were falling objects, people and animals such as rats. The digesters were protected by roofs made from local materials such as thatch, plastic etc. during rain season which could also provide heat and protect the loss of heat to and from the digesters. The roof should be made in such a way that it could be removable during sunny days. Simple fences were made around the digesters to prevent damage by animals or people. The study area was cleaned from shrubs, grasses and unwanted vegetation to protect the damage of rats. Some gas leakage occurred on the joints and simple repairs were made with the help of CN-43 adhesive chemical using vehicle tire inner tube and GI sheet plaster. This result showed that technical problems in the geomembrane digesters were resolved more easily than fixed-dome biogas plant constructed from materials, such as concrete, steel and red mud. Kristoferson and Bokhalders (1991), concludes that in many developing countries, the biogas programs have failed because of high initial cost of investment, inefficient maintenance and lack of technical personnel.

4.13 Environmental Impact Assessment of the Plastic Biodigester

4.13.1 Reduction of green house gas emissions

The environmental benefit obtained from the use of biogas plant as an energy resource is that there is no net production of greenhouse gases. The carbon dioxide released during biogas combustion originally was organic plant material and so is just completing a cycle from atmosphere to plant to animal and back to the atmosphere. Methane is 24 times more severe greenhouse gas than carbon dioxide and capture of biogas as a fuel prevents the release of methane into the atmosphere. Capture of the methane for use as a fuel would significantly reduce the net greenhouse gas production. Additionally, odors are controlled since all the gas is burned prior to release into the atmosphere, pathogens and weed seeds are also destroyed (Guido Gryseels et al., 2007).

According to this study, by the use of one plastic biodigester of size $3m^3$, $971.81m^3$ gas was produced annually from 27375kg cow dung; which could replace the use of 611 litre of kerosene or 3372.2kg of firewood regarding the local price of alternative fuels. This was done by assuming that $1m^3$ of biogas substitutes for 3.47 kg of fire wood, 12.3 kg of dry dung fuel and 0.62 litre of kerosene oil as stated in (UNISCO, 1982; Van Buren, 1974).

The biogas produced from the biodigesters will be used for cooking and lighting and as such consumption of biomass fuel for cooking and consumption of kerosene for lighting will be avoided. Energy per capita consumption of biomass in Ethiopia is 862.4Kg/year (fuel wood=735, crop residue=85, dried cow dung=134, Charcoal=6.4)/year, electricity=28(31) Kwh/year and Petroleum products=21Kg/year which is the lowest in the world (Meskir Tesfaye, 2007).

Thus, the calculation for GHG (CO_2 - equivalent) production per tone of cow-dung in the absence of geomembrane plastic biogas plant installation was done. The average biogas yield of one GPBP from cow-dung under Mersa temperature condition was 35.5 m^3 per tone of cow-dung.

Biogas contains 60% CH_4 and 36% CO_2 (Grewal et al., 2000).Thus, the amount of CO_2 and CH_4 from 1 tone of cow-dung was as follows:

Production of CH_4 / tone = 35.5 * 0.6 = 21.3 m^3

Production of CO_2 / tone = 35.5 * 0.36 = 12.78 m^3

With a conversion factor of 24 for methane (Ernest Lutz, 1998), CO_2 - equivalent of 21.3 m^3 CH_4 was

= 21.3 * 24

= 511.2 m^3

Total production of CO_2 equivalent /tone

= 511.2 m^3 + 12.78 m^3

= 523.98 m^3

With 1.83 Kg/ m^3 as density of CO_2 , total mass equivalent of CO_2 /tone

= 523.98 m^3 *1.83/1000= 0.959 tone

Therefore, 0.959 tone of carbon dioxide (green house gas) which was responsible for the rise in the Earth's temperature and disrupts the heat balance of the Earth was prevented from emitting in to the atmosphere with the use of geomembrane plastic biogas plant.

Moreover, from the use of one GPBP, 360.04 m^3 of CO_2 and 600.06 m^3 CH_4 (14401.44 m^3 CO_2) or 14761.48 m^3 CO_2 was prevented from entering in to the atmosphere and hence decrease global warming. Thus, from the use of 1000 GPBP, 14.76 million m^3 of CO_2 could be prevented from causing global climatic change.

According to ILCA (2007), Ethiopia has a cattle population of 30 Million. When Cow dung from all theses cattle is left to degrade in the open earth, Ethiopia could produce an estimated 6.57 billion m^3 of methane and 2.628 billion m^3 of carbon dioxide per annum.

Thus, this much amount of methane and carbon dioxide has been released in to the atmosphere and increase global warming. Green house gases such as carbon dioxide are responsible for increased global warming. Since biogas is 60% methane and 36% carbon dioxide on the average, combustion of biogas reduces emission of the same gases in the atmosphere. By so doing it contributes to the reduction of global warming.

According to Samuel Faye (2007), smoke from the use of fuel wood and dung for cooking contributes to acute respiratory infections. This problem,i.e., indoor air pollution is worse in poor countries where households' houses are not equipped with separate living and cooking places.Thus,the above problem could be addressed with the use of geomembrane plastic biodigester which produced smokeless and odorless biogas.

According to SURUDE (2002), previously, animal manure was an environmental problem in ThuanAn district of Tanzania, where it caused pollution in the air, water and soil. After installation of the digesters, all 35 families recognized better environmental conditions, less smell, fewer flies and cleaner water. Summarizing details of experiments conducted with pig slurries, pain et al. (1990) cited in SURUDE (2002), concluded that the digestion reduced odour emission by between 70 and 74%. According to the women who were responsible for food preparation, use of biogas meant that they could attend to other work, while cooking .This is in contrast to the situation when using solid fuels such as fire wood requiring much closer supervision. The women stressed that they could now cook in a clean environment free of smoke. Their pots & pans were clean and they did not have to spend time on tedious cleaning. They stated that they could cook all food items on gas (SURUDE, 2002).

4.13.2 Reduction of rate of deforestation

According to EFAP(1999) cited in NWARDO (2007), reported that $10m^3$ of biomass could be obtained from 1hectar of forest and $1m^3$ of biomass is equivalent to 600kg of wood. According to this study, 3372.2kg of fire wood was substituted and saved from the use of $3m^3$ size geomembrane plastic biogas plant which contributes to combat deforestation and soil depletion. Thus, 0.562 hectare of forest per year per GPBP or 562 hectare of forest per year per 1000 GPBP could be protected from deforestation and trees that are not cut down due to wide adoption of biogas technology also contribute to reduction of green house gases in the atmosphere through the process of carbon sink.

According to Mitiku Haile et al.(2006)quoting Drams et al.,2003;Nyssen et al.2003a confirmed that in Ethiopia human induced mismanagement of natural resources is a root cause of soil degradation together with the concomitant climatic changes.

A study on fuel saving biomass stove,"Mirt" in Ethiopia by proved that the use of thousand "Mirt" stoves for baking "injera" for household consumption, it is estimated that 54 hectares of forest will be saved per year from being deforested. So, the use of GPBP has significant contribution than improved biomass stove "Mirt" in minimizing forest area that are being cut for fuel wood consumption.

Similar study was conducted by SURUDE (2002), in Tanzania, showed that each biogas unit is capable of saving up to 37 hectares of wooded land per year. Aggregation of these results can contribute visibly in ensuring environmental sustainability as per millennium development goal.

5 Conclusions and Recommendation

5.1 Conclusions

A field experiment was conducted to introduce economically feasible, technically acceptable and environmentally friendly biogas plant to the farming community and other users in Ethiopia by using four single and double layered geomembrane plastic biodigesters constructed below and above ground surface and one Chinese model fixed- dome biodigester as treatments and using cow dung as an input material. The financial viability of the two models of biogas plants was computed by using cost-benefit analysis. The environmental impact of the new geomembrane plastic biogas plant were assessed and studied by computing the replacement of conventional fuels such as reduction of fuel wood, cow dung and chemical fertilizer with the use of plastic biodigester. The results of the study were summarized and concluded as follows.

Gas production, as the proportion of biodigester liquid volume was higher for a single layered and above ground plastic biodigester than others and very less amount of gas was recorded from the fixed-dome biodigester. This was because on the average more sun light temperature ($32.7^{O}c$) which is useful for fermentation by anaerobic microorganisms was absorbed in a black geomembrane plastic sheeting digester installed above the ground surface than other biodigesters.So, the construction and use of single layered geomembrane plastic biodigester above the ground surface is preferable than other models and locations of installations of the biodigesters. The average temperature measured in the geomembrane plastic biodigesters exceeds 1.7 ^{O}c - 6.7 ^{O}c higher than the fixed-dome biodigester. Thus, geomembrane bag was observed to have the best advantage that the digester contents were heated easily than fixed-dome biodigester made from reinforced concrete. The single layered geomembrane plastic biodigester constructed above ground produced higher total nitrogen than others at a hydraulic retention time of 40 days. This was due to higher amount of slurry temperature absorbed in the single layered above ground biodigester activated anaerobic micro-organisms to convert more simple organic substances of the substrate in to simple organic acid so as to fix more ammonia. Therefore, total-N was affected by material and position of biodigester construction.

Fermented slurry contains larger nitrogen content than fresh cow dung in both models of biodigester.Thus is because, in an air tight biogas digester more organic acid such as acetic acid, prop ionic acid, butyric acid, ethanol and acetone was produced doting anaerobic fermentation of soluble simple organic substances which helps to absorb and fix ammonia and minimize the lose of nitrogen thus conserving the fertility of the manure. Thus fermented slurry high fertilizer value in increasing the fertility of the soil than fresh cow dung.

The digested slurry was thin and used directly to the crops through the irrigation channels and by direct splashing on the farmland. Moreover, it was also put in a fish pond of Mersa agricultural T.V.E.T. College and fishes were nourished. Thus, it was proved that the waste that comes out of the digestion process, as slurry was very useful both as feed and organic fertilizer.

The pH of the slurry was in the range between 6.43 and 6.7. Thus, the condition inside the digestor was very comfortable for aerobic micro-organisms to accomplish normal fermentation and higher gas production.

The geomembrane plastic biodigester gave higher net benefit than the fixed – dome biodigester.Thus, according to the above result obtained from the study, the geomembrane plastic biodigester is the cheapest model that an individual farmer could invest and acquire a better profit than the fixed-dome biodigester.

The social aspect of biogas technology was discussed in terms of the increase income generation through increased crop production and saving biomass fuel expenses with the use of nutritive slurry as organic fertilizer and biogas for cooking so as to solve the problem of fuel shortage in the rural areas when it is popularized at a large scale to the farming community and other users in Ethiopia.

The technology was very simple, sustainable, demand driven and replicable. Main causes of damage to the digesters were falling objects, people and animals such as rats. To alleviate this problem the digesters were protected by roofs made from local materials during rainy season and simple fences were made around the digesters to prevent damage from animals or people, and the study area was cleaned from bushes, grasses and unwanted vegetation to protect the damage of rats. Some gas leakage occurred on the joints and simple repairs were made with the help of CN-43 adhesive chemical using vehicle tire inner tube and GI sheet plaster. This result showed that technical problems with the geomembrane digesters were resolved more easily than with other materials.

Environmental impact assessment of the technology was studied and found that from the use of plastic biodigester, 360.04 m^3 of CO_2 and 600.06 m^3 CH_4 (14401.44 m^3 CO_2) or 14761.48 m^3 CO_2 was prevented from getting into the atmosphere and increase global warming. Moreover, 0.562 hectare of forest could be protected from deforestation per year and trees that are not cut down due to wide adoption of biogas technology also contribute to reduction of green house gases in the atmosphere through the process of carbon sink.

Generally, the geomembrane cylindrical film biodigester technology is cheap and simple way to produce gas and fermented slurry for people in Ethiopia than fixed-dome biogas plant. It could be appealing to rural people because of the low investment, fast pay back, simple technology and positive effects on the environment.

5.2 Recommendations

As there may not be congruence between the time of availability of fresh slurry and time of field application liquid slurry be properly stored or composted. So it is recommended to construct 2 or 3 compost pits closer to the biogas plant so as to conserve nutrients and increase crop yield.

No form of biogas slurry can be profitably left spread on the field. Fields should be ploughed immediately, or if the manure is used as top-dresser, should be covered by soil immediately to conserve nutrients.

Considering the long-term benefit of plastic film biodigester technology both economically and environmentally, it is recommended to introduce the single layered above ground geomembrane plastic biogas technology to be used for the beneficiaries regardless of its higher gas and fermented slurry production via extension education to promote its penetration and diffusion into rural areas. However, greater safety precaution during operation and usage of the plant and protection from damaging agents such as sharpened objects and rats is essential.

Research, development and demonstration are essential activities to overcome the technical obstacles and be well abreast of development in biogas technology. Therefore, similar studies needs to be practiced in the highlands of Ethiopia by involving farmers, creating feedback from the farmers and letting this feedback serve as a foundation for the improvement of the technology.

The government agencies such as Department of Environment, Department of Agriculture and Rural Development, Agriculture Extension Center, and even private institutions could play a leading role by the issue of loans, subsidies and awareness creation extension education services. This could be implemented under government regulations and policies.

References

AoAC, 1990. In: Helrick(Ed), official Methods of Analysis. Association of Official Analytical Chemists 15th edition.Arilington.pp.1230. Retrieved on December 12, 2007 available at www.mekarn.org/msc 2003 -05 / thesis 05 / tram-p, .pdf.

Ayen Mulu, 2004. Amhara National Regional State Head of Government Office. *Agricultural Commodity Marketing System Study Project*, Annex 14, Bahir Dar.

Badege Bishew, 2007. Forest History of Ethiopia Past and Present. Livestock and Pasture Resources. Retrieved on October 4, 2007 available at http:// etff.org/Newsite/articles/article_histryf.html.

Bech, N., Waveren and Evan, 2002. Environmental Support Project (ESP), Component 2:*Environmental Assessment and Sustainable Land Use plan for North Wollo.*

Benjamin Jargstorf, 2004.Renewable *energy and Development Brochure to accompany the mobile exhibition on Renewable energy in Ethiopia.* Addis Ababa.

Biogas support programme, 1994. Nepal *Biogas Plant Construction Manual.* Nepal.

Biogas works, 2000. The history of biogas retrieved on June 20, 2007 from http://www.kursus.kvl.dk/shares/ea/03projects/03projects/32gamle/2005/energy.pdf.
Buixuan An and Preston, TR, 1995. Low cost polyethylene tube biodigesters on small scale farms in China. *Electronic proc.2nd Intl.Conference on Increasing Animal Production with Local Resources*, Zhanjiang, China, p.11. Retrieved on July 15 from http://www.cipav.org.co/Irrd/Irrd 15/7/sant 157.htm

Buixuan An, Thomas TR Preston and Frands Dolberg.(1997, 1999). Livestock Research for Rural Development. *The introducing of low-cost polyethylene tube biodigesters on small scale farms in Vietnam.* Volume 9, Number 2.Retrived on July 15, 2007 from http://www.cipav.org.co/Irrd/Irrd 9/2/an 92.htm

Duong Nguyen Khang and Iminh Tuan, 2002. *Proceeding Biodigester Workshop.*Transferring the low cost plastic film biodigester technology to farmers. Retrived on Sep5 available at http://www.mekarn.org/procbiod/khang2.htm.

EPAP (Ethiopian Forest Action Plan), 1994. Ecosystem Conservation – task force 9.Ethiopian Forestry Action Programme, Ministry of Agriculture, Environmental Protection and Development, Addis Ababa.

Environmental Support Project (ESP), 2001. The Resource of North Wollo Zone,Amhara National Regional State. Thematic reports prepared by DHV ANR BV. Addis Ababa. Ethiopia.

EPLAUA (The Amhara National Regional State Environmental Protection, Administration and Use Authority), 2007. *Simplified Environmental Impact Assessment Guideline.*Bahir Dar.Ernst Lutz, Hans P.Binswnager, Peter Hazell, and

Alexander Mc Calla, 1988. Agriculture and the Environmental Perspectives on Sustainable Rural Development. The World Bank. Washington, D.C.20433, U.S.A.

Ethiopian Agricultural Sample Enumeration, 2002. Results *for Amhara Region*, Part IV, Addis Ababa.

Eye, S. and Yigremew, A, 2000. Farming Assets in North Wollo Statistics, Maps and Impressions from a Travel to North Wollo.

FAO (Food and Agricultural Organization of the United Nations), 1983. Biomass energy Profiles.*FAO Agricultural Services Bulletin 54.*FAO, Rome.

FAO (Food and Agricultural Organization of the United Nations), (1996, 2000). *Statistical data base.*FAO, Rome, Italy.

FaWCDA (Forestry and Wildlife Conservation Development Authority), 1982. FAO/Ethiopia national workshop on fuelwood.FaWCDA, Addis Ababa.

Fischer J R, Iannotti EL, Porter JH and Garcia A, 1979. Producing methane gas from Swine manure in a Pilot-size digester, Transaction of the ASAE.

French D, 1979. The economics of renewable energy systems for developing countries.GTZ, 1989. Biogas plants in Animal Husbandry.

GTZ, 1991. Improved Biogas Unit for developing countries.

Guido Gryseels and Michael R. Goe Highlands, 2007. Energy flows on smallholder farms in the Ethiopian highlands.Retrived on 9/11/2007 from
http://www.ilri.org/InfoServ/Webpub/Fulldocs/Bulletin17/Energy.htm

Gunnerson C G and Stuckey DC, 1986. Anaerobic digestion-Principles and Practices for Biogas Systems. *The World Bank Technical Paper*#49, Washington, D.C., PP93-100.

Gunnerson, 1986; Kristoferson and Bokhalders, 1991; Marchaim, 1992; Karki.1996. Integrated Resource Recovery-Anaerobic digestion, principles .Washington.

Hao et al., 1980; Hong et al., 1979; Umana, 1982. Plastic biodigesters. Retrieved on Sept 16, 2007 available at http://www.fao.org/docrep.to541e/T054/Eog.htm
HARDO (Habru District Agricultural and Development Office). 2006 .Report on Land use and soil type of Habru Wereda.Unpublished

Hu Qichun, 1991.Systematical Study on Biogas Technology Application in Xindu Rural Area, China. Asian Institute of Technology, Bangkok/Thai

ILCA (International Livestock Center for Africa), 2007. Cattle Population of Ethiopia.Retriveon Sep 15, 2007 from http://www.africa.upenn.edu/eue- web/fao-live.htm

IMF (International Monetary Fund), 1982. Confidential *mission report on recent economic developments in Ethiopia*.IMF,Washington,DC.

IUCN, 1990. Ethiopian National Conservation Strategy. Phase 1 Report. Based on the work of M.Stahl and A. Wood. IUCN, Gland, Switzerland

Kristoferson LA and Bokhalders, 1991. Renewable Energy Technologies: Their application in developing countries. Intermediate Technology Publications, London, pp 112-117.

Le Ha Chau.1998a. Biodigester effluent versus manure, from pigs or cattle, as fertilizer for production of cassava foliage (Manihot esculenta).Livestock research for rural development (10) 3: Retrieved on 12 July 2007 from http://cipav.org.co/Irrd/Irrd 10/3/chau 1.htm.

Le Ha Chau.1998b. Biodigester effluent versus manure, from pigs or cattle, as fertilizer for duckweed (Lemna spp.).Livestock Research for Rural Development (10) 3: Retrieved on 12 July 2007 from http://cipav.org.co/Irrd/Irrd10/3/chau 1.htm.

Lekulel, F.P., et al., 2002. *"Technological Interventions for Promoting Smallholder Integrated Farming". A case study of Turiandi, Morogoro, Tanzania.*Retrived on August 20, 2007 from http://sgp.undp.org.

Livestock Research for Rural Development, 1997. The introduction of low-cost Polyethylene tube biodigesters on small scale farms in Vietnam. Volume 9, Number 2.Retrived on August 20, 2007 from http://www.cipav.org.co/Irrd9/2/an92.htm

Mazumdra, 2004. A Text Book of Energy Technology. New Delhi-110 002.

Mitiku Haile, Karl Herweg, Brigitta Stillhardt.2006.Sustainable Land Management-A New Approach to Soil and Water Conservation in Ethiopia. Mekelle, Ethiopia.

MOA (Ministry of Agriculture), 2000. Agro ecological Zones of Ethiopia, Natural Resource Management and Regulatory Department, Addis Ababa

Newcombe K, 1983. An economic justification of rural afforestation: The case of Ethiopia. Draft Energy Department Paper. World Bank, Washington, DC.

N.S.Grewal et al., 2000. Hand Book of Biogas Technology (A practical hand book). Punjab Agricultural University, Ludhiana. India.

NWARDO (North Wollo Agricultural and Rural Development Office), 2007. Household Biomass Fuel consumption .A case study in North Wollo, Woldya. Unpublished.

Paulos Dubale, 2004. *EARO Technical paper*. Addis Ababa.

Pich Sophin and Preston T R, 2001. Effect of processing pig manure in a biodigester as fertilizer input for ponds growing fish in polyculture.Livestock Research for Rural Development. (13) 6. Retrieved on August 10, 2007 available at http://www.cipav.org.co/Irrd/Irrd 13/6/pich136.htm

Preston TR, 1995. Research, extension and training for sustainable farming systems in the tropics. Electronic Proc.2nd Intel. Conference on Increasing Animal Production with Local Resources, Zhanjiang, China, p.3.Retrived on July 17, 2007 available at http://www.mekarn.org/procbiod/Khang.htm

Safely JR, Vetter RL and Smith L D .1987.Managment and operation of a full scale poultry waste digestor.Poultry Science P 941.

Samuel Faye, 2007. Household's consumption pattern and Demand for Energy in Urban Ethiopia.Retrived on 9/11/2007 from http://www.fao.org/DOCREP/004/AB582E/AB582E04.htm

S.C.Santra, 2001. Environnemental Science. New central book agency (p).ltd.India.

Senait Seyoum, 2007. The economics of a biogas digestor.Retrived on Sep 15, 2007 from http://www.ilri.org/infoserv/webpub/Fulldocs/Bulletin30/economi.htm

Stahl, M. and A. Wood, 1989. Ethiopia: National Conservation Strategy, Inception Report. IUCN, Gland.76pp

Stuckey, D.C, 1986. A global perspective. *In biogas technology; Transfer and Diffusion.Ed.M.M.El-Halwagi* Elsevier, London.

SURUDE (Foundation for sustainable Rural Development), 2002. Promotion of low cost biogas technology to resource poor farmers in Tanzania.Retrived on July8,2007 from http://www.equator initiative.net.

Thong Hong Van, 1989. Some experiences on the development and the application on biogas digesters in Dongnai Province.*Proc.First national workshop on biogas application in Vietnam.* Polytechnic Univ.Press, Hochiminh City, pp66-69

Uli Werner, 1989, Nielsen, 2002; Andersen and Sorensen, 2001.The slurry after digestion retrieved on August 10,2007 from http://diggy.ruc.dk(bitstream/1800/363/1/the development of pdf/

UNESCO (United Nations Educational, Scientific and Cultura Organization), 1982. Consolidation of information. *Pilot edition. Genera information programme and universal system for information in science and Technology,* UNESCO, Paris.

UNV, 1983. The use of organic residues in Rural Communities. China. P177.

Van Buren A(Ed), 1974. A *Chinese Biogas Manual: Popularizing technology in the countryside.* Intermediate Technology Publications, London.

Vandana S, 2004. Alternative Energy. APH Publishing Corporation. New Delhi-11002.

Vandana S, 2004. Training course of IREP, organized by planning commission and Gandigram Rural Institute. New Delhi.

Wang Chunlei, 2004.ZIBO *CRANE Plastic Manual.* China.
Werner Kossmann, Uta Ponitz, 2007. Biogas Digest Volume 1. Information and Advisory Service on Appropriate Technology (ISAT). Retrieved on December 13, 2007 from www.gtz.de/de/documente/en-biogas_volume 1 .pdf.

Wikipedia, 2006. Retrieved on Sep3, 2007 available at http://en.wikipedia.org/wiki/Ethiopia.

Yacob Mulugeta, 2000. 'Renewable Energy Technologies and Implementation Mechanisms for Ethiopia', Energy Sources, 2(1).

Appendix
Appendix 1
The questions prepared for interviewing the farming community around the study area
1. **Use of biomass (fire wood, cow dung, crop residue etc)**:- cooking methods, economics of firewood, quantity of firewood used per day, place to be collected, problems faced, solutions suggested by farmers.
2. **Use of other bio fuels (Kerosene oil etc)**:- types of fuels used, cost of fuels, quantity of fuels used per day, problems faced, where to get and solutions suggested by the users.
3. **Availability, operation and utilization of alternative fuels** :- biogas, Electricity, wind mill, water mill, and solar energy

Appendix 2
Calculation of capital recovery factor for Geomembrane Plastic Biodigester
The following standard formula is used to calculate the CRF:

Capital Recovery Factor (CF) = $\dfrac{r(1+r)^n}{(1+r)^n - 1}$

r = discount rate = 18 % = 0.18

N = number of years (estimated economic life of the biogas technology options)

$$CF = \dfrac{0.18(1+0.18)^{10}}{(1+0.18)^{10} - 1} = 0.222515$$

For Geomembrane Plastic Biogas Plant, 10 years service life (Senait, 2007)

Limitations

I have taken the service life of both the double and single layered below and above ground geomembrane plastic biogas plant from literature as 10 years. So detail research would be required on this aspect so as to determine the exact economic feasibility of the technology.

Initial annual investment cost for each type of Geomembrane Plastic Biodigester

PSA = (0.222515)* (1356.15) = 301.764

PDA = (0.222515)* (1917.75) = 426.728

PSU = (0.222515)* (1370.55) = 304.967

PDU = (0.222515)* (1932.75)= 430.066

Operational Cost for each type of geomembrane plastic biodigester is as follows

Digester and gas holder at 5 % depreciation for 10 years i.e.

PSA = 641.00/ 10 = 64.1

PDA = 1143.20/10 = 114.32

PSU = 641.00/ 10 = 64.1

PDU = 1143.20/10 = 114.32

Gas pipeline and appliances at 5% depreciation cost for 20 years

PSA = 310.55/20 = 15.53

PDA = 310.55/20 = 15.53

PSU = 310.55/20 = 15.53

PDU = 310.55/20 = 15.53

Civil Construction Cost at 2.5% depreciation cost for 10 years

PSA = 404/10 = 40.4

PDA = 464/10 = 46.4

PSU = 419/10 = 41.9

PDU = 479/10 = 47.9

Maintenance Cost at 1% of total cost

PSA = 1356.15 *0.01 = 13.56

PDA = 1917.75 *0.01 = 19.17

PSU = 1370.55 * 0.01 = 13.70

PDU = 1932.75 * 0.01 = 19.33

The cost of dung in terms of kerosene oil equivalent

Total fresh dung required for $3m^3$ size biogas plants in one year = 75 x 365

$$= 27375 kg$$
$$= 27.375 \text{ Tone}$$

Quantity of fermented manure measured was 123kg and 112.5 kg for geomembrane plastic and fixed-dome biodigesters.

** Replacement of dung (used as dung cakes) as fuel in terms of kerosene oil equivalent

1Kg of fresh dung = $0.04m^3$ gas. I.e. $1m^3$ gas can be produced from 25Kg fresh dung.

$1m^3$ gas = 12.3Kg of dry dung fuel

Thus, 25Kg fresh dung = 12.3Kg dry dung

And, 27375 Kg fresh dung = 13468.5 Kg dry dung.

$1m^3$ biogas can be replaced by 0.62litres of kerosene oil

And, $1m^3$ gas = 12.3Kg of dry dung = 25Kg of fresh dung = 0.62 litres of Kerosene

Thus, 0.62 liters of Kerosene = 12.3Kg of dry dung

So, replacement of dung used as dung cakes =678.9 liters of Kerosene = 13468.5 Kg of dry dung

1litres of kerosene oil=5.50birr (1999E.C price)

Therefore, 678.9 liters of kerosene = 3733.95 ET.Birr.

So, the total cost of dung in terms of kerosene oil is 3733.95 ET.Birr.

N.B. N and P contents of diammonium phosphate (DAP) roughly approximated the proportions of these nutrients in dried cow dung. Thus 1 tone of DAP is roughly equivalent to 16 tone of dry manure (Senait Seyoum, 2007). The 1999 E.C price of DAP was ET.Birr 1450 tone^{-1}. Thus each tone of dry cow dung is worth EB 90.62 in terms of the plant nutrient it contains.

So, 45 tone of slurry from geomembrane plastic biodigester costs; 45 x 90.62 = EB4077.9

41 tones of slurry from the fixed-dome biodigester costs; 41 x 90.62 = EB 3715.42

The cost of dung in terms of fire wood consumption

Total amount of biogas produced during one year was;

$$= 2.74 m^3 \times 365$$

$$= \underline{\mathbf{1000.1\ m^3 / year}}$$

1m3 biogas = 3.47 kg fire wood

So, 1001.1m^3 biogas = 3473.82 Kg fire wood

1kg of fire wood according to the local price in Mersa was EB 1.00

Thus, 3473.82 Kg of fire wood costs EB 3470.35

Therefore, the cost of dung in terms of fire wood is EB 3470.5

Cost of labor

The cost of one labor in Mesa was EB 15.00

For feeding = EB 342.1875

For transporting slurry = EB 180

Total cost of labor = EB 522.2

Total operational cost for each geomembrane plastic biodigester in terms of manure use

Digester and gas holder + Gas pipeline and appliances + Civil Construction Cost + Maintenance Cost +Cost of dung as manure + Cost of labor

PSA = 64.1 + 15.53 + 40.4 + 13.56 + 4077.9 +522.2   = 4733.69

PDA= 114.32 + 15.53 + 46.4 + 19.17 + 4077.9 + 522.2   = 4795.52

PSU= 64.1+ 15.53 + 41.9 + 13.7 + 4077.9 +522.2   = 5257.53

PDU= 114.32 + 15.53 + 49.9 + 19.33 + 4077.9 +522.2   = 4799.18

Total operational cost for each geomembrane plastic biodigester in terms of fuel use
PSA = 64.1 + 15.53 + 40.4 + 13.56 + 3470.35 +522.2   = **4126.14**
PDA= 114.32 + 15.53 + 46.4 + 19.17 + 3470.35+522.2 = **4177.97**
PSU= 64.1+ 15.53 + 41.9 + 13.7 + 3470.35+522.2   = **4127.78**
PDU= 114.32 + 15.53 + 49.9 + 19.33 + 3470.35+522.2 = **4191.63**

Total cost = Initial annual investment cost + Total operational cost as manure
PSA = 301.764 + 4733.69   = 5035.4532
PDA= 426.728 + 4795.52   = 5222.248
PSU= 304.9674 + 5257.53   = 5562.4674
PDU= 430.066 + 4799.18   = 5229.246

Total cost = Initial annual investment cost + Total operational cost as fuel
PSA = 301.764 + 4126.14= 4427.9032
PDA= 426.728 + 4177.97= 4604.698
PSU= 304.9674 + 4127.78= 4432.7474
PDU= 430.066 + 4191.63= 4621.696

Benefit gained from the different types of geomembrane plastic biogas plant
Therefore, from the above calculations the benefit gained from the geomembrane plastic biodigester is greater than its total cost.

For PDA
From biogas = $2.85 m^3$ * 365 = 1040.25 m^3/year
The price of 1 kg of fire wood was EB 1.00
So with the replacement of fire wood, the benefit acquired was EB 3609.67

For PSA
3.0375 m^3 * 365 = 1108.69 m^3 / year
With the replacement of fire wood, the benefit gained would be =EB 3847.15

For PDU

2.5875 m³ * 365 = 944.44 m³ / year

With the replacement of fire wood, the benefit gained would be = EB 3277.20

For PSU

2.475 m³ * 365 = 903.38 m³ / year

With the replacement of fire wood, the benefit gained would be = EB 3134.71

From Slurry = 123 tones * 365 = 45 tones, this is the same for all plastic biogas plants as per the measured value.

The price of 1 tone of slurry was EB 4077.9

So the benefit from slurry in monetary terms was EB 4077.9

Total benefit = Benefit from biogas + Benefit from slurry

= EB 3470.35 + EB 4077.90

= EB 7548.25

Appendix 3

Calculation of capital recovery factor for fixed dome biogas plant

For Fixed –Dome Biogas Plant, 20 years service life (Yacob,2000)

Initial Annual Investment Cost for Fixed- Dome Biodigester FDB = (0.222515)* (8900) = 1662.698

$$CF = \frac{0.18(1+0.18)^{20}}{(1+0.18)^{20} - 1} = 0.18682$$

For 20 years service life

So, 0.18682 * 8900 = EB 1662.698

Thus, Initial Annual Investment Cost for fixed-dome biodigester was EB 1662.698

Operational Cost for Fixed-Dome Biogas Plant in EB is as follows

Cost of civil construction at 2.5% depreciation for 20 years service life

= 7689.45 /20 = **384.47**

Cost of gas pipeline and appliances at 2.5 % depreciation for 20 years service life

= 310.55/20 = **15.5275**

Cost of maintenance at 0.01% of total investment cost

= 8900 * 0.01 = **89**

Cost of dung as manure and fuel

12.3 Kg dry dung = 3.47 Kg firewood

13468.5 Kg dry dung = 3799.65 Kg firewood

1 Kg firewood = EB 1.00

Thus, 3799.65 Kg firewood costs EB 3799.65

Amount of biogas produced per year was 2.36 m^3/ day*365=861.4 m^3

Amount of fire wood in terms of biogas produced per year = 861.4 m^3*3.47 kgm^{-3}

$$= \underline{2989.06 \text{ Kg fire wood}}$$

And 2989.06 kg of fire wood costs EB 2989.06

Thus, income from biogas in terms of fire wood is EB 2989.06 and,

Income from biogas in terms of slurry is EB 3715.42

Total income from plastic biodigester is EB 6704.478

Total net profit obtained from the use of fixed-dome biodigester in terms of replacement by fuel wood is EB 5073.886 and it's average net profit is EB 2536.943

Cost of labor

    The cost of one labor in M esa was EB 15.00

    For feeding = EB 342.1875

    For transporting slurry = EB 180

    Total cost of labor = EB 522.2

Total Operational Cost for fixed -dome biodigester in terms of manure use

    = Cost of civil construction + Cost of gas pipeline and appliances + Cost

      of maintenance + Cost of dung as manure + Cost of labor

    = 384.47 + 15.5275 + 89 + 3715.42 + 522.2

    = 4726.62

Total Operational Cost for fixed -dome biodigester interms of fuel use

    = 384.47 + 15.5275 + 89 + 2989.06 + 522.2

    = 4000.26

Total cost = Initial annual investment cost + Total operational cost as manure

    = 1662.698 + 4726.62

    = 6389.32

Total cost = Initial annual investment cost + Total operational cost as fuel

    = 1662.698 + 4000.2575

    = 5662.96

Benefit gained from fixed-dome biogas plant

From biogas = 2.36 m^3 * 365 = 861.4 m^3/year

1 m^3 of biogas is equivalent to 3.47 kg of firewood.

Thus, 861.4 m^3 could substitute 2989.06 kg of fire wood

The price of 1 kg of fire wood was EB 1.00

So with the replacement of fire wood, the benefit acquired was EB 2989.06

From Slurry = 112.5tones * 365 = 41 tones

The price of 1 tone of slurry was EB 90.62

So the benefit from slurry in monetary terms was EB 3721.08

Total benefit = Benefit from biogas + Benefit from slurry

$$= EB\ 2989.06 + EB\ 3721.08$$
$$= EB\ 6710.14$$

So, the total benefit is greater than total cost in fixed dome biogas plant.

Appendix 4

Meteorological data for Mersa (Rainfall in mm, from 1981 – 2007 and Temperature in ^0c, from 1994 -2004 and 2007)

Month	JAN	FEB	MAR	APR	MAY	JUN	JUL	AUG	SEP	OCT	NOV	DEC	Total
Rainfall	28.00	87.65	85.75	67.20	28.90	161.55	292.90	159.35	33.1	52.75	31.65	12.00	1040.8
Tmax	25.25	27.40	28.25	30.75	34.46	32.65	29.98	29.85	28.65	27.55	26.95	25.80	
Tmin	12.35	11.20	13.60	14.75	15.14	15.95	15.75	13.62	12.85	11.60	11.15	11.30	
Tave	18.8	19.3	20.92	22.75	24.8	24.3	22.87	21.73	20.75	19.58	19.05	18.55	
Average Tmax = 28.96 ^0c													
Average Tmin = 13.27 ^0c													

Appendix 5

Mersa ATVET College 2006/2007 Metrological data

Month, Year	Max.T,^0c	Min.T,^0c	RF,mm
Sept,2006	28.6	13.4	37.1
Oct,2006	27.8	11.5	87.3

Nov,2006	26.7	13.1	23.6
Dec,2006	25.3	12.9	12
Jan,2007	24.6	13.6	17.2
Feb,2007	27.8	11.2	71.5
Mar,2007	28.7	14.4	70.5
Apr,2007	31.7	15.4	75.7
May,2007	33.51	15.68	24.8
Jun,2007	32.2	16.6	111.2
Jul,2007	29.2	15.9	336.7
Aug,2007	28.95	15.85	222.5
Average T,Oc	**28.755**	**14.1275**	
Total RF,mm			**1090.1**

Appendix 6

Atmospheric Temperature,^{O}c of Mersa during Fermentation of Cow-Dung in Geomembrane Plastic and Fixed-Dome Biogas Plants.

Days	MaxT,^{O}c	Min T,^{O}c
5/5/2007	32.5	15.5
6/5/2007	32.5	15.5
7/5/2007	32.5	16
8/5/2007	32.5	16
9/5/2007	32.7	15.5
10/5/2007	33	15.5
11/5/2007	34.5	15.5
12/5/2007	34	16
13/5/2007	30	16
14/5/2007	33	16
15/5/2007	35.7	16.5
16/5/2007	35	16
17/5/2007	34	15.5

18/5/2007	34.5	17.5
19/5/2007	33.5	17
20/5/2007	34	15
21/5/2007	34	15
22/5/2007	34.5	14.5
23/5/2007	34	15
24/5/2007	33	18
25/5/2007	33	15
26/5/2007	35.5	15
27/5/2007	33.5	15.5
28/5/2007	34	15
29/5/2007	34.5	15.5
30/5/2007	33.5	15
1//6/2007	30	15
2/6/2007	32	15.5
3/6/2007	32.5	15
4/6/2007	33.5	15.5
5/6/2007	33.5	15
6/6/2007	34	15
7/6/2007	34	18
8/6/2007	33.5	15.5
9/6/2007	34.6	17
10/6/2007	33.5	15
11/6/2007	30.5	15
12/6/2007	31	15
13/6/2007	33.5	17
14/6/2007	34	18
Average	33.34	15.75
Total average = 24.54 $^{\circ}$c		

Appendix 7

Slurry Temperature, °c during Fermentation of Cow-Dung in Geomembrane Plastic and Fixed-Dome Biogas Plants.

Temperature Measurement of the Biodigesters using Thermometer, °c.					
Plastic bag digesters & Fixed dome biodigester					
Days	PDA	PSA	PDU	PSU	FBU
5/5/2007	28	30	27	26	23
6/5/2007	29	31	28	26	24
7/5/2007	30	31	29	27	26
8/5/2007	31	33	30	29	28
9/5/2007	30	32	29	28	27
10/5/2007	28	30	28	26	25
11/5/2007	29	33	26	25	24
12/5/2007	32	34	30	29	27
13/5/2007	31.5	33.5	29	27	27.5
14/5/2007	30.5	33	28.5	26	25.5
15/5/2007	31	33.5	29	27.5	26.5
16/5/2007	30	33	30	28	27
17/5/2007	30.5	33.5	30	29	26
18/5/2007	30.5	33.5	30	29	26
19/5/2007	31	34	31	29.5	26.5
20/5/2007	30.5	33.5	30	29	26
21/5/2007	29	33	26	25	24
22/5/2007	29	33	26	25	24
23/5/2007	32	34	29.5	28	27.5
24/5/2007	30	33	30	28	27
25/5/2007	30.5	33.5	30	29	26
26/5/2007	31	34	31	29.5	26.5
27/5/2007	30	33	30	28	27
28/5/2007	28	30	28	26	25
29/5/2007	28	30.5	26.5	25	22
30/5/2007	29	31	28	26	24

1//6/2007	30	33	30	28	27
2/6/2007	30	33	30	28	27
3/6/2007	30.5	33.5	30	29	26
4/6/2007	30.5	33.5	30	29	26
5/6/2007	30	33	30	28	27
6/6/2007	30.5	33.5	30	29	26
7/6/2007	30	33	30	28	27
8/6/2007	28.5	31	28.5	27	25.5
9/6/2007	29	31.5	29	27.5	26
10/6/2007	30.5	33.5	30	29	26
11/6/2007	30.5	33.5	30	29	26
12/6/2007	30	33	30	28	27
13/6/2007	30.5	33.5	30	28	
14/6/2007	30	33	30	28	27
Average	**29.988**	**32.7**	**29.175**	**27.65**	**25.963**

Appendix 8

Geomembrane Adhesives

CARMYFIX CM-43 is an excellent general purpose synthetic rubber adhesive that occupies a leading position in the shoe and furniture industry for the last decades. It is a top standard neoprene adhesive that offers strong and long-lasting adhesion for leather, upholstery, paper, elastic, cloth, wood, metal and many other materials. It is made in Greece.

To attach geomembrane plastics, the CM-43 adhesive should be mixed with other chemical mixtures with the only geomembrane plastic technician working as energy and mines expert in North Wollo, Gubalafto district Agriculture and Rural Development Office.

Appendix 9

Description of Biogas Double Delux Burner

A biogas deluxe burner which has a capacity of 450 liters/hr for each burner could be used for cooking purposes. It consists of 1. Nozzle, for air and gas inlet, 2. Mixing chamber, to hold air and gas mixture 3. Fire sieve element, for the passage of flame outside 4.Valve to allow the passage of gas to the mixing chamber 5.Air shutter used for nozzle adjustment to get the desired flame temperature.

3

1

2

4

5

Appendix 10

Model of Geomembrane Plastic Biogas Plant

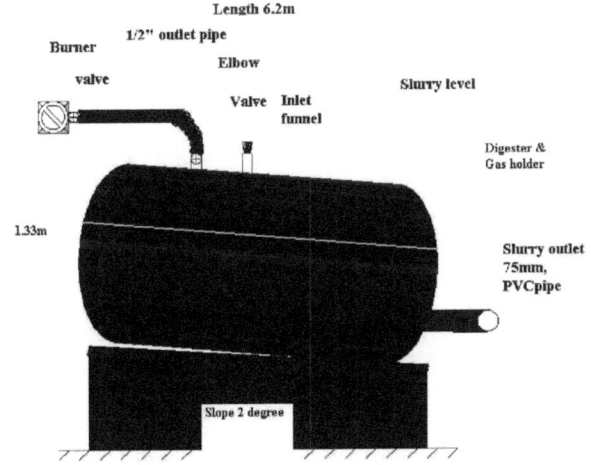

Appendix 11

Description of the Biogas Digester

The geomembrane plastic biogas digester has 6 major components (Appendix 18):

DIGESTER AND GAS HOLDER: It is a cylindrical plastic bag about 6 m long and 1.33 m in diameter. The bag consists of 1 and 2 layers of 5mm black geomembrane plastic sheets used to store the influent and effluent.

INPUT INLET: made from 3/4 " GI socket and 3/4" GI nipple with GI cape used for delivering input materials in to the digestor.

OUTLET: consists of 1.5 m long and 80 mm diameter wide PVC pipe, 3/4" GI socket and 3/4" GI nipple. It is employed for slurry outlet.

GAS OUTLET: made from 3m long and 1/2" diameter wide PVC pipe, 3/4" GI socket, 3/4" GI nipple, reducer and 1.5m long neoprene rubber hose used for the passage of gas to the point of utilization.

ACCESSORIES: An Indian model cooking burner of capacity 450 liter/hr. used to burn and measure the amount of gas generated from the biodigester and 3/4" gate valve to regulate the gas flow.

List of Tables

Table 1: Work Load before and after Biogas Production 10
Table 2: Quantity of cattle dung Required for Feeding of different Sizes of Bio-gas units 12
Table 3: Potential gas Production from Different Feedstock 12
Table 4: Land use in Habru Woreda 20
Table 5: Technological Parameters of the Experimental Biodigesters. 22
Table 6: Amount of Cow-Dung and Water Fed to the Biodigesters 24
Table 7: Total Values for Biogas Production (9 Am to 4 Pm) in Bio-digesters with different Types, Layers and Location of Installation. 30
Table 8: Comparison of Average Slurry Temperature, ^{O}C and amount of Gas Produced, m^3/Day of the Biodigesters 31
Table 9: Effect of Material & Position of Biodigester Construction on the Composition of the Effluent. 35
Table 10: Summary of Market Value of Inputs and Outputs Used in the Analysis 39
Table 11: Initial Cost of Investment for Geomembrane Plastic Biodigester in EB. 41
Table 12: Operating Cost for Geomembrane Plastic Biogas Plant in EB per year. 42
Table 13: Summary of total annual discounted costs for geomembrane plants in EB per year 43
Table 14: Benefit Obtained from the Geomembrane Plastic Biodigesters in EB per year. 44
Table 15: Initial Cost Investment for Fixed-Dome Biodigester 44
Table 16: Operating Costs for Fixed-Dome Biogas Plant in EB per year. 44
Table 17: Summary of total discounted cost for fixed-dome biogas plant in EB per year. 45
Table 18: Benefit Obtained from Fixed-Dome Biogas Plant 45
Table 19: Summary of total costs and total benefits of the two model biogas plants 45

List of figures

Figure 1: Ethiopia's Primary Energy Shares, Benjamin Jargstorf (2004) 6
Figure 2: Fuel Wood Carriers for Fuel Consumption from the Forest, Agent of Deforestation, Benjamin Jargstorf (2004) 7
Figure 3: Dry Cattle Dung Cakes used for Fuel in Ethiopia, Benjamin Jargstorf (2004). 7
Figure 4: Geomembrane Plastic used for Biodigester Construction 17
Figure 5: Geomembrane Plastic Welding Machine,(photo shoot by Author) 18
Figure 6: Map of the Study Area 19

Figure 7: Layout of the Experimental Site 22
Figure 8: Field Installation of the Plastic Biodigester after Feeding (Photo taken by the author) 23
Figure 9: Plastic Biodigester at the Beginning of Gas Generation (Photo taken by the author)... 29
Figure 10: Burning and Measuring of Biogas with a Biogas Burner after Gas Generation (Photo taken by the author) 30
Figure 11: The Relationship between Temperature of the Slurry and amount of Gas Production for the Plastic and Fixed-Dome Biodigesters. 31
Figure 12: Comparison of Total-N Fresh Cow Dung & Fermented Slurry for Plastic and Fixed Dome Bio-digesters. 36
Figure 13: Comparison of Organic Matter (Kg) Content of Fresh Cow Dung and Fermented Slurry for Plastic and Fixed Dome Biodigesters.. 37

Acronyms

A.T.V.E.T.	Agricultural Technical Vocational and Educational Training
BOD	Biological Oxygen Demand
CED	Mekelle University (Center for Environment and Development, Bern University)
COD	Chemical Oxygen Demand
CSA	Central Statistics Authority
C:N	Carbon to Nitrogen Ratio of the substrate
CM-43	Geo-membrane Plastic Camry Fix Adhesive
DAP	Di-Ammonium-Phosphate
DM	Dry Matter
EB	Ethiopian Birr
EIA	Environmental impact assessment
EREDPC	Ethiopian Rural Energy Promotion and Development Center
EPLAUA	Environmental Protection, Land Administration and Use Authority
ESP	Environmental Support Programme
EARO	Ethiopian Agricultural Research Organization
EFAP	Ethiopian Forest Action Plan
FAO	Food and Agriculture Organization of the United Nations
FaWCDA	Forestry and Wildlife Conservation Development Authority
FBUE	Effluent from Fixed-Dome Biogas Plant
FDU	Fixed-dome bio-digester constructed underground surface
FFB	Fresh cow dung fed to the fixed-dome biogas plant
FPB	Fresh cow dung fed to the geo-membrane plastic biogas plant
GARDO	Gubalafto Woreda Agricultural and Rural Development Office
GDP	Gross Domestic Product
GPBP	Geo-membrane plastic biogas plant
GPWM	Geo-membrane Plastic Welding Machine
GI	Galvanized Iron
GTZ	Deutsche Gesellschaft fur Technische Zusammenar beit
HARDO	Habru Woreda Agriculture and Rural Development Office
HDPE	High Density Polyethylene Geo-membrane
HRT	Hydraulic Retention Time
ILCA	International Livestock Center for Africa

IMF	International Monitory Fund Organization
IUCN	The World Conservation Union
IFB	Influent fed to the fixed-dome bio-digester
IPB	Influent fed to the geo-membrane plastic bio-digester
ISO	International Organization for Standardization
LDPE	Low Density Polyethylene Geo-membrane Plastic
Mg/Kg	Milli gram per kilo gram
MOA	Ministry of Agriculture
NGO	Non-Governmental Organization
NPV	Net Present Value
NWARDO	North Wollo Agricultural and Rural Development Office
^{o}C	Degree Celsius
ODM	Organic Dry Matter
PDA	Plastic Biodigester, double layered and constructed above ground surface
PDAE	Effluent from Double Layered above Ground Geo-membrane Plastic Bio-digester
PDU	Plastic Biodigester, double layered and constructed underground surface
PDUE	Effluent from Double Layered under Ground Geomembrane Plastic Biodigester
PSA	Plastic Biodigester, single layered and constructed above ground surface
PSU	Plastic Biodigester, single layered and constructed under ground surface
PSAE	Effluent from Single Layered above Ground Geomembrane Plastic Biodigester
PSUE	Effluent from Single Layered under Ground Geomembrane Plastic Biodigester
PVC	Poly-Vinyl Chloride
SURUDE	Foundation for sustainable Rural Development
TLU, %	Total Livestock Unit in percent
UNISCO	United Nations Educational, Scientific and Cultural Organization
UNV	United Nations Volunteers
VS	Volatile Solids